The Gift of Touch

HELEN COLTON

KENSINGTON BOOKS

are published by

Kensington Publishing Corp.
850 Third Avenue
New York, NY 10022

The author gratefully acknowledges permission from the following sources to quote from material in their control:

Harry N. Abrams, Inc., for material from *Manwatching: A Field Guide to Human Behavior*, by Desmond Morris. Text copyright © 1977 by Desmond Morris.

The American Humanist Association for material from "Why Sport?" by Ed Cowan, published in *The Humanist Magazine*, November/December 1979. Copyright © 1979 by The American Humanist Association.

Arbor House for material from *The David Kopay Story—An Extraordinary Self-Revelation*, by David Kopay and Perry Deane Young. Copyright © 1977 by David Kopay and Perry Deane Young.

Brigham Young University, Center for International and Area Studies, for material from *Culturgrams*. Copyright © 1979, 1981 Center for International and Area Studies.

Doubleday & Company, Inc., for excerpts from "Planet Pass" and "The Lap Game," from *The New Games Book* edited by Andrew Fluegelman. Copyright © 1976 by The Headlands Press, Inc. And for excerpts from *The Hidden Dimension*, by Edward T. Hall. Copyright © 1966 by Edward T. Hall.

W.H. Freeman and Company for material from article "The Anthropology of Manners," by Edward T. Hall, published in *Scientific American* Magazine, April 1955. Copyright © 1955 by Scientific American.

Macmillan Publishing Co., Inc., for material from *Even the Night*, by Raymond L. Goldman. Copyright 1947 by Raymond Leslie Goldman.

W.W. Norton and Company, Inc., for material from *Anatomy of an Illness*. Copyright © 1979 by Norman Cousins.

Public Broadcasting Communications for material from "Touching and Being Touched," by Robert Coles, published in *The Dial*, December 1980. Copyright © 1980 by *The Dial*.

Scholastic Inc. for material from "Please Touch! How to Combat Skin Hunger in Our Schools," by Dr. Sidney B. Simon, published in *Scholastic Teacher*, October 1974. Copyright © 1974 by Scholastic Inc.

World Future Society for material from "Future Fun—Tomorrow's Sports and Games," by Lane Jennings, published in *The Futurist*, December 1979. Copyright © 1979 World Future Society, 4916 St. Elmo Avenue, Bethesda, Maryland 20814.

Ziff-Davis Publishing Company, for material from article, "How Cultures Collide," by Edward T. Hall, published in *Psychology Today*, July 1976. Copyright © 1976 by Ziff-Davis.

First Zebra Books Printing: 1989

First Kensington Trade Paperback Printing: February, 1996

ISBN 1-57566-012-1

Printed in the United States of America

ACKNOWLEDGMENTS

I am grateful to:

Susan Bolotin, for her wisdom and clarity of thinking.

All the librarians at the Los Angeles Public Library—especially my neighborhood branch library—for their enthusiastic energy and help in locating books for me.

Ward and Barbara Tabler for their exquisite precision in the use of language and for their clippings.

Ruth and Jim Stewart for their clippings, notes and loving words.

Irv Good for listening so much to my talking about writing this book.

To my sisters and brothers
Rae, Sylvia, Leatty, Ben and, in memoriam, Irv

And to my friends
Dina Mellon and Goldie Kahn

CONTENTS

THE GIFT OF TOUCH

WE KNOW IT NOW, MORE THAN EVER IN THIS ALIENATED AND HOSTILE SOCIETY, WHERE THE ONLY REAL COMMUNICATION LIES IN THESE PRIMAL ACTS OF LOVE AND TOUCH—MAN AND WOMAN, MOTHER AND CHILD, FATHER AND CHILD, HUMAN AND HUMAN. WHAT ELSE ENDURES?

—MARYA MANNES,
OUT OF MY TIME
(DOUBLEDAY, 1971)

INTRODUCTION

My first awareness of the power of touch came during the interview with Marilyn Monroe described in Chapter 3, "Touching in Family Life." Ever since then I've had the gnawing thought, Why was that touch so important to Marilyn?

A long time later, I heard a talk by Dr. Solon Samuels at the USC School of Medicine on the René Spitz study of infant mortality described in Chapter 2, "Being Born—Our First Touch Experience." Like the interview with Marilyn, this too was often in my mind.

Then, about 1975, Dr. Elmer Belt, an admirer of my books and articles, sent me a gift copy of Ashley Montagu's pioneering book *Touching—The Human Significance of the Skin*, about an infant's tactile needs during its first year or two. What about tactile needs throughout our lives? I wondered.

I had to undergo three more experiences before I felt that final impetus to get started on a book on this huge subject. One, I read James Prescott's study, "Body Pleasure and the Origins of Violence," mentioned in "Being Born—Our First Touch Experience." Two, I attended a conference convened by Laura Huxley on Project Caress, mentioned in the first chapter. And, three, I came across a map of the human brain that showed that the largest area on our brain surface is devoted to the hand. That fascinating map gave me the "Eureka! I have found it!" feeling that I needed to get started.

Researching and writing this book has changed my life. I am more loving, more joyous, less fearful, more at peace. I hope reading it will change your life too.

Helen Colton
West Hollywood, California, 1983.

1

Touch—Our Lost Sense

Right now, as you're holding this book, what are you experiencing with your sense of touch? Observe what your fingers and hands are feeling. Are your fingertips touching the binding of the book or the book jacket? Is the binding smooth or rough? Is the book jacket sleek and slippery, printed on coated paper? Are the four fingers of each hand holding the book open while your thumbs rest on the page you are now reading and the page opposite it? Is the paper pleasurable to your touch? Does it feel soft or harsh? Run your fingertips lightly up and down the page. Is it possible to feel any slight bumps or indentations of the ink?

Riffle the pages with your thumb. Is it pleasurable for you—as it is for me—to sense the pages rapidly slipping away from under your thumb?

Perhaps the book feels heavy on your fingers and is setting up muscle tension in your fingers and thumbs, pressure on the palms of your hands?

You may be reading this book while you are riding a bus to work on a freezing-cold morning. You are wearing woolen gloves and clumsily trying to turn pages with your thickly covered fingers. Or it may be summer and you are lying on a webbed chaise longue beside a swimming pool. You become aware that the crisscross of the mesh is digging uncomfortably into your skin and leaving marks on it.

Or you may be feeling lonely, longing to be cuddled in someone's arms, yearning for the warmth and heartbeat and solace of another human body. Poet-composer Rod McKuen says, "The need to touch someone can be so great at times that it's as close to madness as I ever hope to come."[1] Lacking such close contact with a person, you may be deriving pleasure and inner serenity from stroking a beloved pet.

Touch is the most important, and yet the most neglected, of our senses. We can survive without sight; blind people do. We can survive without hearing; deaf people do. We can survive without being able to taste; many of us do. We can survive without our sense of smell; smell is the first of our senses to leave us when we fall asleep and the last to return when we awaken—that's why people often cannot smell a fire when they are asleep. Our anthropoid ancestors and the earliest of humans, walking on all fours, had their noses close to the ground to sniff the odors that alerted them to danger. Now, as we climb up the evolutionary ladder, our sense of smell is slowly atrophying.

But we cannot survive and live with any degree of comfort and mental health when we are not able to *feel.* A complete loss of our sense of touch can send us into psychotic breakdown.

A study of boredom was done at McGill University in which male subjects lay on beds in lighted cubicles twenty-four hours a day. They wore translucent plastic visors that permitted light to reach their eyes, but the objects they saw were blurred. They lay with their heads on U-shaped pillows that limited but did not cut off their ability to hear. On their hands they wore cotton gloves and cardbord cuffs extending beyond their fingertips so that at no time were they able to feel anything with their fingers and hands. In a short time—only a matter of hours for some—they were unable to think clearly. Their thoughts became disjointed and incoherent. They thought in fragments of sentences, then forgot in the middle of a sentence what they were thinking. They experienced numbness, lethargy, a free-floating sensation of their physical beings gradually disintegrating.[2]

Even a partial and temporary loss of touch can be terrifying. One evening Jacqueline Du Pré, the noted cellist, was preparing to give a concert, when she suddenly lost sensation in her hands. She opened the case in which she kept her cello and experienced a shiver of horror; she could not feel the case. She took the bow in her hands and could not feel the bow. Throughout the concert, her terror grew as she felt no sensation in her hands and fingers. "I managed to find the notes by gauging every position of my fingers with my eyes," she said later. Subsequently, she was found to have multiple sclerosis, a disease in which touch signals are not transmitted by the central nervous system to the brain.

Anthropologist Richard Gill, through an accident, temporarily lost his sense of touch. He said, "I had not realized before how entirely dependent our fineness and accuracy of movement are upon the sense of touch. Our walking and movement in the dark, our most instinctive reachings-out or sittings-down or hand-graspings, are all but impossible if we can't feel the objects we are touching. I was horrified all the time until my sense of touch returned."[3]

Every one of us is born with intense skin hunger. At birth, we derive all sensations and information through the largest organ of our bodies—our skin. The tactile sensations that we receive when we are snuggled up to a warm body, feeling the rhythm of a beating heart as we suckle at a breast or drink from a bottle are vital to our survival. They set up electrical impulses that travel along our neural pathways and help create biochemical reactions that enable our brains to develop and function.

Some of our touch needs are met—but not enough, as we shall see in Chapter 2, "Being Born"—when we are fed, diapered, bathed and carried as babies. But when we start getting around on our own, touching too often lessens or ends. Many of us go through life touch-starved. We suffer malnutrition of the senses and rarely experience that glorious feeling of having our touch hunger fully satisfied.

Hunger is an apt analogy. Each of us is a combination of many of the 106 chemical elements currently recognized by science. These elements are contained in the chemicals produced in our brains. Touch stimulates the production of chemicals in the brain, and these feed our blood, muscle, tissue, nerve cells, glands, hormones, organs. Deprived of touch to stimulate these chemicals, we may be as starved as if we were deprived of food.

It is ironic that our society, which is uncomfortable with touch and usually does not encourage it, certainly encourages it when the profit motive is involved. Underneath many commercials on television and advertisements in magazines is an appeal to our sense of pleasure through touch. Toilet tissue commercials talk about how "soft, softer, and softest" they are. A man's cheek is more pleasant to touch when he uses a particular brand of shaving cream or lotion. Hair is smoother to touch when it is washed with the advertiser's product. But let people talk about touching for humanitarian reasons—just to enhance understanding and good feelings between people—and we consider them bizarre, hippie or oversexed. When the human-potential movement started on the West Coast in the 1960s at such places as Esalen Institute, where touching and hugging were encouraged, the rest of the country ridiculed California's "touchy-feelies." The fact is, we can learn from them.

We have a long way to go to become comfortable with touching. Gene Shalit, a host on the morning TV show *Today*, is in the habit of touching co-host Jane Pauley's chair and occasionally her arm or shoulder. A viewer, uncomfortable with even this scant contact, wrote a protest, to which Pauley commented that, in determining the public's acceptance of touch, it was difficult to know when "a pat on the back is too little and a pat on the upholstery is too much."[4]

Another indication of how far we have yet to go to understand our touch needs are the hundreds of books on family life, many of them used as texts in schools, that do not have *a single mention* of touch. Many books on communication skills also never mention the word *touch*. Even more incredibly, the word does not appear in the indexes of several sex encyclopedias. How can we teach young people the basic requirements of good family and marital relationships when we ignore our most vital sense?

Nowhere in conferences devoted to bettering family life, such as the White House Conference on Families, or the American Family Forum sponsored by General Mills Corporation, is there a mention of what family members feel about touch, how they touch, what kind of touching they would like to have from each other, and what they feel when touch is missing in family life.

As we shall see in Chapter 3, "Touching in Family Life," families live for years in intimate quarters, yet suffer skin starvation. Undeniably, this lack of touch contributes to our high rate of divorce and family upheaval. Bill Johns, a Los Angeles therapist who deals with teen-age runaways, says he believes "ninety percent of the girls who run away are partly motivated by guilt and shame over their longing to be touched and held by their fathers."

We *talk out* but do not *act out* our need for touch. Listen to these common expressions:

Of a sentimental or emotional occasion we say, "It was a touching experience."

We tell an inspiring speaker, "Your talk touched me deeply."

The poet says, "There is a touch of spring in the air."

Of a kindly or generous person we say, "She's a soft touch."

"Don't be so touchy," we tell a grouchy person. (In fact, if a grouch *were* a touchy person—touching others or being touched—he undoubtedly would be less grouchy.)

Recipes tell us to use a "touch" of oregano or basil or thyme.

"I'll 'touch it up' with paint," we say of a messy surface.

Some of us, unfortunately, may suffer a "touch" of rheumatism.

A ship "touches" port when it makes a brief stop.

One of the most frequently spoken sentences in the land is probably "Let's keep in touch," said as we part from a friend, relative, colleague, or acquaintance. How much less alienated we would be if, instead of just mouthing that phrase, we did actually touch. I did exactly that with a woman I met at a workshop. We touched thumb to thumb, a declaration that we were kindred souls in this human race together. We enjoyed our touch signal and, as we parted, we both started laughing joyously. Life felt good!

No one of us is without a need for contact with a warm being. Somerset Maugham wrote, "I do not like being touched and I have always to make a slight effort not to draw away when someone links his arm in mine."[5] Such people, like the rest of us, were born with skin needs, but they may have experienced a birth trauma, a childhood injury or a physical punishment that makes them cringe when they are stroked. Some people are so strongly indoctrinated with a moral code that equates affectionate touching with sexual arousal that they convince themselves they do not need or enjoy skin closeness.

Touch is a great social leveler reaching across class, economic, color and ethnic boundaries. I used to be intimidated by authority figures, but not anymore. Talking to an austere financial tycoon, a mover and shaker of nations and economic systems, I picture him going home, dropping his emotional armor, and laying his head like a child on the breast of a nurturing wife or lover, having her cuddle away the tensions of his day.

Touch is also the most social of our senses. We can see, taste, smell and hear by and for ourselves, but touch is the one sense requiring us to interact with others—at least most of the time.

In a higher sense, touch gives us all kinship. There is something Godlike about the power every one of us possesses in our hands and fingers to bring pleasure and give meaning to another human being. We can all be creators of "social enzymes" that add up to well-being. Like Walt Whitman, writing in "Song of Myself," we can say: "I make holy whatever I touch or am touch'd from."

On a recent visit to the hairdresser, the woman who did my hair looked drawn and haggard, rushing through her appointments, being interrupted by phone calls. I observed this while I was under the dryer. Just before she was to do my comb-out, I told her to sit down and I rubbed her back, neck and shoulders. She closed her eyes and leaned back gratefully. With a few moments of gentle massaging and kneading, the tense lines in her face disappeared. Her whole being went limp with relaxation. I kept my hands gently on her shoulders just to transmit my warmth to her. The woman who got up from that chair was entirely different in appearance and demeanor from the person she had been before she experienced my touch.

My touch was not only a physical communication; it also held psychological meaning. I'm sure my hairdresser sensed that. It said that I was aware of the stress that she was feeling, and not because she was just someone I was paying for a service but because she meant something to me as a person, I wanted to comfort her and ease her fatigue.

The one area in which we give ourselves permission to enjoy touching is in our contact with animals. Zoologist Desmond Morris says, "Pets are not for looking at, for studying, or for admiring at a distance. They are for fondling and cuddling, thereby replacing intimacies that are missing from the ordinary lives of pet-owners." He analyzed photographs of owners and their animal friends. Over fifty percent were holding their cats, dogs, rabbits, hamsters, even birds, in their arms as if the pets were babies. Five percent were kissing their pets, mouth to mouth.[6]

Why do we have a sense of touch at all? What is Nature's intention in giving us millions of sense receptors throughout our bodies and embedded in our skin? Nature is our greatest security guard. Above all else, our sense of touch alerts us to danger—through temperature, vibration and pressure. Nature wants the human race to survive.

Your fingers or hands touch a hot pot or stove, and without thinking, your reflexes draw you away from danger. Put your hand into a bucket of ice and you will also draw your hand away in an involuntary action to escape the pain. Your sense receptors find these extreme temperatures unsafe for survival.

A friend was asleep under an electric blanket when a sudden pain from heat around her feet awakened her. An instant later the blanket burst into flames. That warning through her sense of touch enabled her to save herself.

You go to climb a ladder and as you put your hands on its two sides, delicate vibrations tell your sense receptors that it is dangerously wobbly. You're driving your car and you apply the brakes. Instantly you feel nuances of vibration on the sole of your foot, warning you that your brakes are not holding with their customary tightness.

I was especially grateful for my sense of touch recently when I dashed out of bed to answer the doorbell. Receptors on my ankle instantly alerted me, by the lightest of pressures, that I had caught my foot in the telephone cord. I steadied myself in time to prevent a headlong fall against the wall.

Among Nature's intentions in giving us touch is to inform and educate us, to help us differentiate surfaces and motions. Run your fingertips lightly across surfaces near you right now. Your sense of touch is capable of discriminating between a smooth pane of glass and etched glass; between sandpaper and slick paper; between a china plate and a paper plate. Your fine discrimination can tell whether you are rolling a grain of rice or a grain of sand between your fingers. There are 1,300 nerve endings per square inch in your fingertips. You don't even have to touch with your whole fingertip; you often get information by touching just with your fingernail. We can get information from locations a sixteenth of an inch apart on our skin. That's why, if you are lying on your stomach on the beach, you can feel that a watermelon seed is stuck on your back, or an insect is crawling on you.

Frank A. Geldard of Princeton University's Cutaneous Communication Laboratory, a pioneer in developing a new language, "vibratese," in which blind people will be able to read by print being translated into vibrations on their fingers, reports that the skin can even handle time almost as well as the ear does. "Under proper conditions, the skin can detect a break of about ten thousandths of a second in a steady mechanical pressure or tactile buzz."[7]

All these sensations help us to sort out and give meaning to the millions of inputs we constantly are receiving. Life would be maddeningly chaotic if we did not possess such tactile discernment.

When I put my car into cruise control while driving the

freeway, I cannot tell from *seeing* my car if the cruise feature is engaged. I know it only when I *feel* it by the car's motion. I like feeling the car's sudden forward surge because I enjoy driving without having to keep my foot on the gas pedal.

One day, as my hand was turning the key in the front lock of my home, I knew at once from the motion of the lock that someone had been inside. The cylinder did not turn with its usual tension but revolved loosely. Sure enough, a burglar had been there and had departed by the front door.

When your pet wants to get your attention, she will nuzzle back and forth against your leg or ankle. Your sense of touch understands—"I want to go for a walk"; or "Pick me up and hold me"; or "Come, look at something I want to show you."

Touch is so magnificently efficient that even the degree of pressure informs us. Deep pressure stimulates different nerve endings than does light pressure. An airy touch often says something affectionate, nurturing or sexual. A heavy touch to the same part of the body may be a warning, "Cut it out. Stop doing that." Someone gives your cheek a light nip or pat, and you feel it as playfulness, an expression that the person admires and cares for you. But that same nip, with hard pressure, means something completely different. Television producer Norman Lear told Phil Donahue that when his dad would give him a pinch on the cheek as a kid, he knew it was meant as no loving gesture but as an angry reprimand.

A husband who makes a sexual overture to his wife often cannot understand why she becomes annoyed rather than amorous. A man puts too heavy pressure into a "love pat" on his wife's behind. She feels it as an unpleasant sting, and her adrenaline pumps with anger. He has prepared her body not for lovemaking but for "fight or flight."

You probably have had this common experience in which the degree of pressure gives you a message about a stranger. You are sitting in a theater next to a stranger. Your elbows touch lightly as you use the armrest at the same time. If the stranger's elbow were to touch yours with hard pressure, it would seem to be a hostile jab, and you might become annoyed or angry, give the person a dirty look, and avoid any more contact with this selfish, rude person whose touch is saying, "I'm not going to share this armrest with you."

I recently learned something very useful about my own degree of pressure. I was carrying a letter to the mailbox when I suddenly realized that my fingers and palm were clutching that quarter-ounce envelope hard enough to carry many pounds. Why on earth am I carrying this with so much force? I wondered. Then I realized that I had transmitted to the object in my hand my anxiety about the response to my letter. As soon as I recognized that, I lightened my pressure on the envelope. Instantly, my whole body relaxed.

Our touch pressure is a powerful indicator of our stress at any given moment. Talking on the telephone, do you grip the receiver with great tension? Washing a pot, are you using as much pressure as you would need to scrape the paint off a piece of furniture? Become aware, as I did, of how much unnecessary force you are constantly exerting through a day, transmitting with the intensity of your touch the intensity of your feelings about what you are doing. By lightening your pressure, you help relieve your stress. Take an envelope and try it.

We often express stress by clenching our upper and lower teeth. Clench your teeth hard; observe how tense you feel. Now open your mouth so that your teeth are slightly apart. By releasing the pressure of your clenched teeth, you relax your whole body.

Nature is not only our guard and educator, she is also a wily seductress. The same receptors that alert and educate us also lure us into sex. Delicate play over our bodies transmits delicious messages from our skin to our brains to our gonads. The most exquisite pleasure the human brain knows comes through sexual orgasm. This shows what a shrewd manipulator Nature is as she says, in effect, "I'll give you your greatest pleasure; you give me the greatest number of humans." To ensure our addiction to sex, Nature has made the most sensitive areas of our bodies, next to our fingertips, lips and tongues, the clitoris and the penis.

About a third of our five million receptors are in our hands, the most sensitive receptors being concentrated in our palms and fingertips. To test how highly sensitive your fingertips and palms are, put this book down for a moment. Lightly stroke the fingertips of one hand against the fingertips of the other hand. You may feel a pleasurable stirring in your genital area, perhaps a tickling sensation, perhaps a feeling of well-being. Your mood may become lighter and more joyful, your body may feel relaxed. Now stroke the palm of one hand with the index finger of the other. Continue doing so for a few moments; you may find yourself gasping, as I do, with the eroticism of that simple touch.

The next-most-sensitive areas are our lips and tongues. To test your lips' sensitivity, pluck a hair from your head. Pull it back and forth across your upper and lower lips. The receptors are so exquisitely attuned that the fine hair, infinitesimal in its circumference, can feel like twine. A particle of food caught between your teeth can feel like a boulder, and your tongue keeps trying to clean it out.

When our sense receptors receive the stimuli of touch, what happens? The touch sparks a minor volt of electricity that shoots into a neuron. You have a hundred billion neurons, or nerve cells. Each is separated from every other one by a junction called a synapse. As your electrical charge goes through a neuron, it stimulates a chemical neurotransmitter, which sparks across that synapse and ignites another electrical impulse through the next neuron, which does the same for the next neuron, and so on. It is similar to the Morse code. Your touch or someone's touch on you is the lever that beeps the message. Your brain is the receiver.

Your brain receives the message in its pleasure center if you are being enjoyably caressed, embraced or stroked. It receives the message in a pain center if you step on a tack, stub your toe or accidentally bite your tongue. (Our tongues hurt so much when we accidentally bite them, because they have some of our most sensitive receptors.) Your brain also receives the message in its executive, fact-absorbing or fact-analyzing, part that tells you the texture, temperature and shape of what you are feeling—whether it is rough or smooth, hard or soft, cold or hot, a light pat or a hard pressure, whether an object is curved, pointed, round, rectangular, flat like a playing card or cube-shaped like dice.

While all this goes on you are playing an inner biological symphony as each pop, crackle—actually minuscule explosions—of your neurotransmitters emits sounds. An imaginative composer, listening to amplified sounds of our bodily processes, could probably create a Heartbeat Concerto, Colon Cantata or Nervous System Symphony!

Reaching your brain, the electrical impulses trigger the manufacture of neurochemicals. The human brain is the greatest pharmaceutical manufacturer in the world. It creates chemicals and combinations of chemicals that can do nearly everything done by synthetic chemicals that we take in the form of drugs. In fact, the pharmaceutical industry constantly works to create drugs that "mimic" in their effects the chemicals we produce in our brains. These chemicals, combining with hormones and enzymes in our blood, give us our moods, sensations and feelings.

Thus, if you are feeling blue as a result of chemicals in your bloodstream, and you are cuddled by a warm, caring person, that touch can result in your brain discharging chemicals that counteract the chemicals creating your sad mood and make you feel better. This happens in as short a time as one millionth of a second. According to Manfred Eigen, director of the Max Planck Institute for Biophysical Chemistry in Göttingen, Germany, that's how long it takes for our brains to have a chemical reaction to sensory input like touch.[8]

Stop now and look at those incredible parts of yourself—your hands. A map of the brain shows that your hands occupy the largest territory in your cortex, next to the territory alloted to your face and mouth.[9] The digit occupying most space on that area of your brain is your thumb. Although it is shorter than your fingers, it is your most versatile. It can go across your palm, it can wrap around your other fingers folded together, and it can enscribe the biggest circle. It is our only digit that is "apposable"—it can touch every other finger facing it. In this one fact, according to anthropologists, lies the beginning of *Homo sapiens.* When they developed an apposable thumb, our simian ancestors could grasp objects. This led to their being able to hold a slab of rock with one hand and bang away with a slab in the other hand to fashion primitive tools, beginning a new phase of evolution.

Move each of your fingers up and down and circle it around. You probably have never realized before that your thumb and index finger have the most mobility, your pinky (its name derived from the Dutch word, *pinkje*) comes next, and your ring and middle fingers are able to make the smallest circles. Your ring finger may have become that for two reasons: wedding rings were traditionally placed on this finger of the left hand because it was thought to be the link to the organ of love—the heart—on our left sides. It may also have been favored for adornment because it is less mobile and therefore less utilitarian than our other fingers.

Fingers and hands constantly give out messages, telling us that we are liked or loved, bawling us out, stroking us, soothing us, angering us, apologizing for us, congratulating us, telling others how upset or ecstatic we are. Frank Costello, despite his imperturbable manner while testifying before a

Senate committee investigating gangland activities, mesmerized the nation when television close-ups of his fingers showed them constantly twisting, twisting, twisting, revealing his inner agitation.

As a counselor I may need to give negative feedback to clients. To lessen their pain on hearing something negative about themselves, I pull up a child's chair to the sofa where the client is seated, take his or her hand and arm and stroke it in a loving way. I am saying in skin talk, "Even though you do this negative thing, you are still a good person. I care about you. I want you to care about yourself."

In expressing loving feelings, a touch says more than words do. Let's say you are with a lover. The two of you see something that delights you. You take each other's hand and give what actress Liv Ullmann calls "secret squeezing signals" that say, "I'm sharing this with you, my love. This is something the two of us have together that nobody else is part of."[10]

When lovers quarrel, their first action toward reconciliation is frequently a touch. The warmth and pressure of their skin starts healing their wounds before conversation does. Meta Carpenter Wilde, a Hollywood script girl who had a love affair with William Faulkner, tells how they made up after an estrangement by speaking "the language of a man's hand enclasping a woman's. The gentle pressures, the exploration of fingers, and finally the peace of two hands fitted together."[11]

Our bodies digest the loving meaning from such touch as if it were health-giving food. It is likely that the constant "up" feeling, the high when we're in love, is the result of having a steady supply of tender squeeze and touch signals. It is the absence of skin talk that makes our lives so barren when we are without love.

When I counsel couples, I ask if they feel loved by each other. I am not surprised when a wife says, "No, I don't really feel loved by my husband." Frequently he is astonished, and he exclaims "But I often tell you I love you."

Saying "I love you" is a combination of sounds spoken into the air. *Touching* loved ones records your emotions on their bodies; they *feel* your love. A wife says that she gets weak-kneed with joy when her husband, helping her with her coat, squeezes her shoulder. Another woman told a researcher that "I would rather be held by my husband every day—though not twenty-four hours a day—than have a Cadillac convertible."

Desmond Morris says, "A single intimate body contact will do more than all the beautiful words in the dictionary. The ability that physical feelings have to transmit emotional feelings is truly astonishing."[12]

Why are we becoming aware of our awesome sense of touch? In the past, touching has been associated mostly with sex. Sexuality was severely monitored by governments and religion, which set up laws sanctioning intimacy only within marriage and mostly for procreation. Touching was considered part of sex, to be reserved primarily for procreative experiences. Now our awareness that the earth is a crowded tenement with more people than it can provide for has led to better population control, with the result that we are no longer allowing ancient laws and religious injunctions to control our bodies. For the first time in history we are insisting upon owning our own bodies. Owning ourselves, we are becoming a pro-pleasure human race. Among the pleasures that we are discovering is the joy of touching—how sweet and satisfying it can be to communicate our meanings to one another. In his book *The Hidden Dimension*, Edward T. Hall states that touch gives us "the velvet quality of satisfaction."[13]

Another reason for our beginning to care about touch is that, as medical costs soar, we want to learn skills for healing ourselves so that we can lessen our need for drugs and institutionalized medical attention. High on the priorities of the holistic health movement is touch, encouraged by practitioners who are successfully using "hands-on" techniques such as massage, acupressure and hugs, to bring relief to aching bodies and minds.

Researching human physiology, we are learning that when the pleasure centers in our brains are turned on, our pain centers are turned off. By providing more pleasure to our bodies with touch, can we lessen the duration and intensity of pain?

The women's-equality movement is another factor stimulating our interest in touch. There is throughout much of the world a strong double standard that says it is okay for men to touch in many situations and relationships but not okay for women to do so. Freeing "the politics of touch" is among the goals of the feminist movement.

Our interest in touch is also a reaction to the overkill of the electronic age. We are rebelling against mechanization that inflicts hard, cold, unyielding machines on our lives, making us long for the feel of soft, warm, yielding flesh.

Finally, our desire for touch is a reaction to the brutalized, harsh, depersonalized conditions of modern life that make so many of us angry, depressed, often violent, people. Tired of the dog-eat-dog lives that many of us lead, fearing one another, our spiritual natures yearn to be loving toward others. We can get closer to an ideal physiological state called *homeostasis*, when all our parts—hormones, electrical energies, enzymes, muscles, bones, tissue, blood—are acting in harmony. The constant striving of the human organism is to reach this state of homeostasis. Touching helps us to get there.

There are many indications of our changing attitudes toward touch. The World Council of Churches sponsors touch workshops at annual meetings. Some churches have rock masses, which close with parishioners and visitors clasping hands or hugging to the beat of joyous music. During a rock mass at an Episcopalian church in Pasadena, strangers swept me into their arms and danced me around the sanctuary. At a Worldwide Marriage Encounter, 15,000 people in the religion-based movement crowded together on the campus of the University of Southern California to create "the world's biggest hug." Phyllis Barber reported in the *Utah Holiday Magazine* in 1980 that Spencer W. Kimball, head of the Mormon Church, "often greets both men and women with a frontal bear hug and is unusually open and affectionate in the way he touches others."[14]

Judges and juries are awarding financial settlements to persons who have lost some of their sense of touch due to accidents. A woman was awarded $15,700 because an automobile accident left her without any sensation of touch in her lower lip and she could no longer enjoy kissing. A Los Angeles judge has ruled that minimum-security prisoners awaiting trial must be allowed to have physical contact with their visitors.[15]

Medical schools are beginning to include courses for doctors and nurses on when and how to touch patients. Hospitals throughout the country now ask for volunteers to come in and cuddle "failure-to-thrive" babies. Foretelling our possible future, author Laura Huxley, in the forefront of a national campaign to raise our consciousness about the vital need for touching, is developing Project Caress. She says:

> We envisage in every city block a serene, soundproof, pastelcolored room, furnished with comfortable rocking armchairs and pillows. In this room, adults would give an hour of their time to HOLD A BABY, knowing that their warmth and affection will magically infuse its entire life with responsive tenderness.

At the moment of writing this in longhand on a yellow pad, I am sitting at a counter in a coffee shop near the Los Angeles Public Library. It is early on a Saturday morning, and I am going to spend the day researching the subject of touch. I decide to do an experiment on myself, to become aware of as many sensations as I can during my breakfast.

I have ordered coffee and a toasted English muffin. It is a chilly morning. The coffee shop is not heated. I curl my fingers around the hot mug of coffee; the sensation is absolutely delicious, as if the temperature were actually a nutrient feeding me. The warmth from my fingers spreads through my body. It is like magic.

I raise the warm muffin to my lips, and I feel the melted butter oozing from it. That restimulates childhood sensations of biting into warm cake fresh from the oven and not caring about the esthetics of eating neatly. I open a little square plastic container of orange marmalade and spread it on my muffin. Some of the marmalade gets onto my left index finger. It feels sticky. I lick the stickiness off with a dab of my warm, moist tongue.

I pretend for a few moments that I have lost my sense of touch and I cannot feel the knife with which I am spreading the marmalade. I cannot feel the cup as I raise coffee to my lips. Almost instantly I become surprisingly disoriented and physically uncomfortable, as though I had no roots on the planet and were floating in space. It is like being in horrible suspended animation, not to be able to feel. It is a physical relief to stop playing my little game and reassume my sense of touch.

I am aware of the mug on my lips, the wet liquid trickling down my throat, the texture of the English muffin against my lips and tongue and in my mouth. It is a coarse texture, not pleasant. I become aware of the reassuring feel of the pencil in my hand as I make these notes.

I take my paper napkin, dip a corner of it into my glass of cold water and wipe my mouth, noting all the while the feel of the napkin (rough, coarse, cheap paper), the cold water (unpleasant; I would prefer warm water on this cold day), and finally the absolute pleasure of wiping the uncomfortable stickiness from my mouth.

Suddenly I become aware of the feel of my eyeglasses on my ears, on the bridge of my nose, and across my upper cheeks. My bottom becomes aware that it is sitting on a hard wooden stool, not padded. I feel a growing excitement at discovering my sense of touch. A whole new world is opening up to me. My sense of touch gives me a comforting and reassuring rootedness on this earth. I feel as if I can never again be bored, because I have found my sense of touch!

As I leave the coffee shop, I note the feel of my vinyl wallet as I take out money to pay my check, the cold metal of the counter as I lay down my money, the leather of my purse as I adjust its strap on my shoulder. I walk down the street feeling the touch of my feet making contact with the pavement, the comfort of my toes inside my round-toed oxfords. There suddenly comes into my mind a memory of the Russian woman Rosa Kuleshova, whose sense of touch was so acute that she could tell the colors of papers by touching them with her fingertips. I long to have such special magic in my fingertips.

I am ecstatic about my discoveries. I am creating a serenade of the senses. In an odd way I feel as if, at this instant, I am for the first time fully, thoroughly, completely alive. I feel the sensation of warmth and wetness as tears well up in my eyes and roll down my cheeks. I feel almost exalted, as if I am in an altered state of consciousness and having a peak experience. I want to share this joy with everybody on earth.

Now I know what Rousseau meant when he wrote, "To live is not merely to breathe; it is to act, it is to make use of our organs, senses, faculties, of all those parts of ourselves that give us the feeling of existence."

2

Being Born—Our First Touch Experience

Imagine yourself, lying head down, in your mother's uterus. You are warm and snug as you gently float in the amniotic fluid, feeling deliciously weightless, protected by layers of fat, tissue and skin between you and the outer world. You cannot smell or taste. After about six months in your cozy nest, if your mother were standing in a powerful light or bright sunshine, you might perceive with your primitive sense of sight a golden haze surrounding your dark home. If a sudden loud noise such as an explosion occurred, you might experience the "startle reflex"—your miniature body would jump in response to the noise.

But you can *feel*—and you feel all the time. You feel the rhythmic *thud, thud, thud* of your mother's heartbeat, accompanied by the *slosh, slosh, slosh* of blood racing through your umbilical cord bringing you oxygen, proteins and other nutrients. As your mother walks, her rolling and swaying give you an enjoyable sensation of being rocked. Occasionally you do a fetal dance, stretching your arms and legs as you push against the inner wall of her womb. That feels good to you—and to your mother, who exclaims, "I just felt the baby kick."

All during your long adventure of becoming a human being, it is your sense of touch that sustains you, gives you information, comforts you, feeds you and pleases you. No wonder that, as early as 1894, the *American Journal of Psychology* called our sense of touch "the mother of senses."[1]

After about 266½ days of growing in your mother's uterus—that is approximately 6,400 hours that it takes us to be created from conception to birth—something miraculous happens. It may be that the final centimeter of your 450-centimeter brain has been created. Perhaps your one hundred-millionth neuron or your sixteen-billionth brain cell has suddenly developed. Or your final eyelash has sprouted. Science does not yet know what happens to start the birth process at the instant it begins. Doctors believed something happened in the mother's body to trigger the birth process, but research is revealing that it is the baby, not the mother, who decides when to start being born. Dr. Graham Liggins, an obstetrician-gynecologist at the University of Auckland, reported at the 1975 Annual Conference of the Obstetrical and Gynecological Assembly of Southern California that the fetus, at the precise instant at which its hypothalamus tells it that its organ systems are mature enough to go it alone in the world, releases a chemical that triggers the release of other chemicals in the mother and thus induces labor.[2]

At that instant you start the most momentous journey of your life—the journey down the birth canal. Your voyage will be hilly and smooth, slippery and tight, dark, confusing and painful as you are squeezed hard by your mother's contractions and then forced forward through this nightmare tunnel as she bears down. In the uterus you could move about, now you are crushed into a tight chamber, your arms and legs tightly locked against your body as you are pushed and shoved.

Suddenly, shockingly, after those months in that dark, warm, soft, safe place, and those hours on your terrifying journey, your head emerges. And how does the world welcome you? With a frightening bombardment of your senses. You see floodlights and the startling white garments of doctor and nurses. Your eyes hurt from the glare. You hear loud noises of voices. You feel hands lifting you into a huge space and turning you upside down. You are panicky—you feel as if you are going to fall. The fear of falling is one of the fears we are born with. Suddenly you feel a sharp pain as the doctor slaps you. Sharp cold metal scissors cut your cord, just as you are in the midst of a throb of nourishment from it. You are placed, naked, on a cold scale to be weighed. A light shines into your eyes as a stinging liquid is dropped into them. Rubberized hands and rough cloth wipe the protective cheeselike coating from your body, hurting your delicate thin skin.

Exhausted from the grim, hellish voyage, you scream and flail about, grimacing with terror, thrashing your arms and legs wildly into space, disoriented after the confined and protected weightlessness that you have known. Now there is no wall or barrier for your body to push against; there is only open space. What a horrible experience being born is!

Margaret Mead described this method of birth as "skin shock." The shock is worsened because your umbilical cord is cut while it is still pulsating, you are not offered food, and you are put alone into a bed in a room loud with the shrieks of other unfortunates like yourself. Swaddled tightly in a straitjacket of a blanket, you can't even move your arms to comfort yourself by being able to put your thumb into your mouth and suck on it. You are desperate for the floating, touching, rocking sensations that you had in the warm womb. And so you cry. And scream. And sob your protest.

This birth process may explain why many of us, even when things are going well in our lives, often awaken with heavy, depressed feelings, almost of dread. We have no logical explanation for why we feel this way. Waking up has been described as feeling like being born. It is likely that each awakening restimulates in us the sadness of our first waking—being born.

While writing this chapter the meaning of a lifelong recurring dream became clear to me. I have dreamed of trying to crawl through a tiny suffocating tunnel. There is an underground aspect to my dream as if I were doing something illegal, like digging a tunnel under a prison, smuggling out something, or hiding. I feel claustrophobic, frightened, as I try to escape, crawling on my belly, my face close to the ground. I always awaken before I crawl all the way through. I suddenly realized, with a chill down my spine, that this was what struggling to be born was like, suffering as I tried to reach the opening at the end of the tunnel.

Dreams of birth trauma are common. Hannah Tillich, widow of theologian Paul Tillich, wrote: "I dreamed somebody was trying to choke me. I must have been in the grip of my own elbow, having folded my arm too tightly around my neck. The dream might have had to do with my having been nearly strangled in my umbilical cord at birth."[3]

If you too may be having unpleasant recurrent dreams reliving your birth, these words may help to bring that dream from your unconscious to your conscious mind and exorcise it. That is what has happened to me; since writing this, I have never again had that dream.

Concerned about the harm done to human beings by the birth trauma, a French obstetrician-gynecologist, Frederic Leboyer, seeks to change the way children are now delivered. During several decades, Dr. Leboyer delivered some nine thousand babies in France. He became increasingly distraught as he observed terror, anger and pain in the eyes of newborns and felt their bodies frantically thrashing about in his hands. Fighting the delivery procedures of hospitals, ridiculed by medical people who disagreed with him that babies suffered during birth, the pioneering doctor began to deliver babies by the gentle method he worked out, first imagining himself going through the birth channel, experiencing what it must have felt like for every infant to emerge with such terror in its eyes. "If our deliberate intention was to teach the child that it had fallen into an indifferent world, a world of ignorance, cruelty and folly, what better course of action could we have chosen?" he asks.

Dr. Leboyer seeks to make the newborn's first minutes of life a harmonious continuation of the comforting touch experiences it has known during its more-than-six-thousand hours in the womb, to reunite mother and child instantly after birth with what he calls "the language of lovers—touch." This takes more time than the traditional method, which enables a doctor to be in and out of the delivery room in an hour or so, arriving in time for the birth—after nurses have performed the preliminary procedures—and leaving as soon as he can decently make his exit.

A friend gave birth to a brain-damaged child, after the nurse, cooperating with a tardy doctor who didn't want to leave his golf game on a Sunday afternoon, pressed the infant's head back into the birth channel so that the doctor could be there for its emergence. The parents sued and received a settlement from the hospital and the doctor.

Our American speed-up ethos dictates that whatever takes longer—including the delivery of human life—cannot be better. That is one reason American hospitals and doctors are reluctant to adopt Dr. Leboyer's methods. Obviously it does take longer for the physician to wait several minutes for the infant's cord to stop pulsating before cutting it than it does to snip the cord instantly, which is the current practice. It certainly takes longer for the physician to place the infant, belly down, on the mother's abdomen and give it a slow, gentle, deep massage for several minutes, alternately following one hand with the other hand down the length of the infant's spine and body, to re-create the sensation of the ebb and flow of uterine fluid.

It takes more time and energy, too, for the physician to engage in other Leboyer procedures. After the long massage, he bathes the infant in a bathinette with water the same temperature as amniotic fluid. Slowly he rocks the child back and forth with its body submerged, to simulate the motion of the womb. The baby decides by its body responses how long this bath is to be, anywhere from ten to twenty minutes. Only when its tiny body is relaxed, free of jerkings and flailings, is it removed from its postnatal hydrotherapy. The infant is then gently dried, dressed in a warmed diaper (have you ever experienced the delicious feel of a warmed towel after you step out of a bath or shower?) then is wrapped loosely in a cotton gown and given back to the mother for more touch and heartbeat sensations.

Babies delivered this way have been photographed within minutes of birth, and they look alert, intelligent and happy, their beatific expressions evidence of joyful feelings they are having about being handled unhurriedly and lovingly, with concern for their touch and motion needs.

Because skin hunger is poorly fed during our first hours, we are frantically hungry for touch feedings. To thrive, newborns must be fed touch as much as food. Dr. René Spitz, working at a hospital for abandoned babies and babies whose mothers were in prison, became alarmed that even though the infants were well-fed and kept in highly sanitary conditions, they suffered a high death rate from a disease called *marasmus*, a Greek word meaning "shriveling-up or wasting-away of the flesh without apparent medical cause." Vacationing in Mexico, Spitz visited an orphanage in which conditions were less sanitary but the babies were happier, more robust and alert, and cried less. He observed that women from the village came in every day to rock and fondle the babies, talking and singing to them. Subsequently he observed thousands of babies. Touched babies thrived, while those who were left alone in bassinets tended to become ill, their cells dying of touch starvation.[4]

Testing the effects of touch on premature babies, researchers came up with similar findings. At the University of South Carolina Medical School, it was found that "preemies" who received four fifteen-minute periods of stimulation each day gained weight and grew faster than unstimulated preemies, and with fewer feedings.[5] A nursing student at Emory University, Mary McFall Jankovic, cast new light on an old theory that the less a preemie was handled, the better off it would be. Keeping track of twelve preemies, she showed that those who were "held and stroked showed more signs of relaxation after feedings, had a decrease in postfeeding pulse rate, respiratory rate, muscle-tension rate, neck hypertension and crying behavior."[6]

In a sense we are all born preemies. For some reason that we don't yet understand, Nature has us born half done after gestation of about 266 days—making us the only species helpless so long after birth. Birds quickly fly, colts run, fish swim, reptiles crawl, but we need about nine more months before we can get around on our own muscle power and crawl or toddle. We finish gestating *outside* the womb. To do this, our brains desperately need "somato-sensory" stimulation so that new cells can be born to help us to grow and complete our primitive nervous systems. We crave temperature, touch and rocking motion similar to that which we enjoyed *inside* the womb.

Babies who are fed enough of the nourishing brain food of touch may crawl, walk and talk at early ages and may develop higher I.Q.s than those not sufficiently touched. The University of Virginia Medical School explored the effect of touching and rocking on ninety-two children. Those whose parents reported a great deal of early stimulation performed "significantly higher on language, memory, learning of new information, and visual-spatial problem solving."[7] Children whose sense receptors make them feel deeply loved find it easy to perform well. (The same applies to adults.)

Tactile stimulation not only may increase infant intelligence, it can also lessen the effects of birth injury. Shirley and Charles Robinson of Los Angeles have thrilling evidence of this. In 1979 Shirley, after thirty-six hours in hard labor, gave birth to a son. "When Sterling was about six weeks old," she says, "I was stunned to observe that there was a great difference between him and other babies his age. There was a striking difference in the use of his eyes. *Sterling wasn't looking at me or at anything else!*"

Tests showed that the infant was not responding to light flashed in front of his eyes or following the light with eye movements. He had suffered a vein rupture, a stroke, while being born. He was diagnosed as "a spastic child with cortical blindness." An opthalmologist and a neurologist agreed that Sterling would never see.

When Sterling was three months old he began to have seizures, which terrified Shirley. Drugs were prescribed. Shirley, who believes in health foods and no medication, was distraught about giving phenobarbitol to her infant. Unwilling to accept the negative prognosis, she constantly sought programs that might help her son. One day she visited Margaret Martin, director of the Pregnancy and Natural Childbirth Center, who had just completed a course at the Institute for the Achievement of Human Potential in Philadelphia, a place noted for helping brain-damaged children. Shirley found there was a waiting period of a year in Philadelphia. She says, "If we waited that long Sterling would lose precious time. Every day that a brain-injured child isn't getting treatment that can help him get better, he is probably getting worse."

Then the Robinsons had a stroke of luck. Charles heard a radio announcement for the Help for Brain-Injured Children Foundation in La Habra, California.[8] There Sterling was given tests, and a program was designed to stimulate his brain. Known as "patterning," the exercises require three people, one person at the child's head, one person at each side. In rapid, synchronized motions, all three move the head, arms and legs in circles, up and down, from side to side. Little Sterling was also spun in a revolving chair, rolled on the floor, massaged, squeezed, tapped and tickled. "The whole process was seasoned with love and a positive attitude," reports Shirley. One month after they started the touch-and-motion program, Sterling could see! Today, despite the doctors' predictions, he has full vision, and while he does not yet have complete muscular coordination for his age level, he is making progress toward crawling and walking. Shirley, Charles and friends, grateful for what they have learned about touch stimulation, have organized Mother-Child Enrichment Classes to pass on their knowledge to others.[9]

In Scandinavian countries, when an infant is ill, parents try "skin therapy" before seeking medication. Liv Ullmann tells of this experience with her daughter: "Linn is four weeks old. She has colic and cries and cries. Ingmar [the child's father] sits with her in bed. He undresses her—then himself; places her tiny body, stiff with cramps, against his bare stomach. She quiets down and in each other's warmth they fall asleep."[10]

What might happen when babies who are not fed enough touch food? Some scientists believe that babies with undernourished, hungry brain circuits deprived of pleasurable nutrients may become predisposed toward violence. A neuropsychologist, James W. Prescott, formerly with the National Institute of Child Health and Human Development of the United States Department of Health, Education and Welfare, makes a startling statement: "The principal cause of human violence is a lack of bodily pleasure derived from touching and stroking during the formative periods of life." Studying forty-nine societies from the past and present, Prescott found "strong support linking physical violence in a person's adult life to lack of physical affection when he was a child." He concludes: "Those cultures that give a great deal of infant physical affection—a lot of touching, holding and carrying—were rated low in adult physical violence. Conversely, the cultures that rated low on physical affection of children were rated high on adult physical violence."[11]

To see how touch-deprived babies might become disposed to violence, let us go back to being infants in the days following birth. You do not know words. You experience sensations only on your skin and in your mouth. Warm, breathing "blobs" or "things" feed and diaper you. You do not know that that "thing" is a mommy or a person. All you care about is that you get enough food and touch.

If you are not handled and rocked enough, you may lie there lonely, starved for warmth and bodily contact. You cry. Nothing happens. You do not have muscle power to help yourself. You feel horribly helpless. Eventually you stop crying, because it doesn't bring any response. For whatever reason—perhaps those around you have the archaic belief that a crying child is "spoiled" by being picked up, perhaps because the mother has too many children and not enough energy to hold and caress all of them, perhaps because those around you just don't have the information about how vitally you need cuddling and rocking—your brain is seriously undernourished even though your stomach might be full.

You are filled with rage—at yourself for being so unworthy and unloveable; at the whole world for treating you so cruelly. As you grow up, anger and hatred are often with you, sometimes on the surface, sometimes hidden under a layer of resignation or passivity so others won't know how mad you really are. It is easy to hate others when you hate yourself first. You distrust people, because people have not proved trustworthy to fill your needs. You secretly enjoy lashing out with a punch, a pinch, a slap, a kick, a blow. Now you have the power to hit back in protest against the people who did not give you what your skin and brain required in infancy.

Noted studies by Harry and Margaret Harlow at the University of Wisconsin showed that baby monkeys deprived of their mothers' bodily comfort grew up to be irritable, aggressive, snarling and violent. Human babies show this same behavior. Some grow up seeming always to be on the hair-trigger edge of violence. We have known such people. When we are in their company we feel uneasy as we figuratively walk on eggshells lest we crack their fragile egos and set off their explosive tempers. We may think, What a short fuse he has.

None of us can afford to be indifferent to the correlation between a lack of satisfying closeness and a potential for violence. Touch-starved—another way of saying *love-starved*—children "are likely to become bondless, hollow men and women who contribute largely to the criminal population," says child-rearing authority Selma Fraiberg.

> Their non-attachment leaves a void in that area of personality where conscience can be. Where there are no human attachments there can be no conscience. The potential for violence is far greater among these bondless men and women. We must look upon a baby deprived of human partners as a baby in deadly peril. This is a baby who is being robbed of his humanity.[12]

This could explain the sadistic violence committed in a New Jersey community. For years the town maintained a Children's Barnyard, a place of delight where everyone could fondle chickens, ducks, rabbits, pigs, turkeys, geese, sheep, goats, puppies and ponies. Several mornings in a row, custodians found ducks with their legs broken, chickens strangled, geese with their eyes gouged out, rabbits stomped to death, a pigeon smashed to death with stones. All the perpetrators turned out to be young children. Perhaps, on an unconscious level, these youngsters were envious of pets who received what the children wanted for themselves—holding and caressing.

Acts of violence excite human beings, as participants or observers. We experience elevations in blood pressure, heartbeat and breathing rate. We like, occasionally, to feel our hearts racing and our pulses pounding. Such bodily experiences provide us with needed stimulation. Psychiatrist Helen A. DeRosis writes in *The Family Coordinator* of October 1971 that "it is not the violence per se but a sense of aliveness which, with his capacity and need for feeling, man is always seeking."[13] Some people who don't get stimulation through joyful contact seek it through harsh contact. A teen-ager just out of jail was asked on the TV show *Today* why he believed he had gotten into trouble with the law; he answered, "I'm one of those people who needed to experience something."[14]

Certainly, lack of touching does not account entirely for violence-prone personalities. Many of us grew up with less physical nourishment than we would have liked. I did, as one of six children with harassed, overworked parents. And yet we are not violent. Other factors too are at work.

Conditions that may tend to dispose us to violence include poor nutrition in the mother's womb and in early life while bones, brain and nervous system are forming, due to lack of information and the lack of essential protein foods; overcrowding; being beaten or seeing beatings; living in a materialistic society whose credo is: "Profits and possessions above people" and where violence, as in war, is a source of profits; watching mass media that repeatedly condition us to view punches, guns and knives as power sources that provide solutions to problems—all of these may dispose us to violence. Yet the majority of children reared in such conditions do not become criminals. Many become academic and professional achievers.

Dr. Samuel L. Woodard of Howard University spent a year in close contact with twenty-three junior-high students in

Washington who lacked at least one parent, who lived in ghetto housing, whose families had poverty-level income, and who performed brilliantly at school. Despite their material impoverishment, Dr. Woodard revealed in a personal conversation, the students all reported that they *felt* loved by one or both parents. A major difference between youthful criminals and achievers was in the positive stroking, physical and psychological, that parents gave children.[15]

A loving touch costs nothing. When you cuddle a child, she doesn't know whether she was born into a rich or into a poor family. At that moment she possesses the world's greatest riches—the protection she feels as your arms enfold her and the bodily message that you love her.

What can loving, nurturing people do to help every child receive its birthright—the joy of touching? Even if we do not have small children in our lives, each of us, as a matter of self-interest, must encourage a new childrearing philosophy in which society takes responsibility for providing pleasurable touch experiences for all babies, beginning at birth. We must end two widespread childrearing notions.

One: "You spoil a crying child by picking it up." Children cry mostly because of unmet needs, for cuddling, socializing, wanting to be warm, dry and fed. Children are "spoiled"—their tender psyches are damaged—when their needs are not met, not when their needs are too well met. Consider how it is for you as an adult. When your physiological and psychological needs are unmet, you are irritable and angry. When needs are met, you have feelings of well-being; you find it easy to be loving and kindly to others.

Scientist Jane van Lawick-Goodall decided to raise her son "with a great deal of physical contact, affection and play," as she had seen chimpanzee mothers raise their offspring, "and never to leave him screaming in his crib." She later reported that in spite of what some societies might consider "spoiling" or indulgence, her son grew up to be extremely independent, sociable and caring for others.[16]

Two: "Spare the rod and spoil the child." The United States has the world's highest incidence of child beatings and murders by parents. Criminologist Marvin E. Wolfgang reported at a professional conference that the street violence that causes so many of us to live with constant fear is a direct extension of the violence that parents inflict on children.[17] UCLA psychiatrist Louis Jolyon West, an authority on violence, has never studied "a murderer of the heinous or wanton type who was not himself a victim of truly terrible violence in childhood."[18]

I hurt physically when I see a parent hitting a crying child at the supermarket. Sometimes I say sympathetically, "It's hard to shop with tired children. Do you think you could get someone to stay with your child next time so you can shop by yourself?" The parent is ashamed to continue the beating, knowing that someone is watching. It also expresses an awareness of the parent's harried plight.

(Whether or not a stranger has the right to speak out about a parent mistreating a child is a philosophic matter that plagues most of us. We are torn between, on the one hand, our belief in rugged individualism and the right to privacy, even if that includes beating children, and on the other hand, our gut response that makes us feel that it is our human and social duty to protest. I have no such ambivalence. I like the attitude of the Chinese, who believe that one has an *obligation* to society to protest mistreatment of children.)

Next to providing warmth, pleasure and comfort, the greatest gift you can give a small child is a home in which the child can safely explore. Babyproof a home by removing valuable and breakable objects during the year or two of a child's explorations. Living for a while without favorite objects is a small price to pay for having a child who finds learning fun because its earliest explorings were approved. Constantly saying "Don't touch" to a child becomes translated into "Don't be curious. Don't explore. We don't like it." Many grownups have lost life's precious sense of adventure and curiosity because their early explorations were inhibited with too many admonitions not to touch.

Stimulating a child's appetite for touching can save parents a common distress—marketing with a child who reaches for items on shelves and badgers parents into buying something. Adults often misread a child's intent. He does not necessarily want to *own* every item; he wants to explore it by touching it—exactly what he sees grownups doing.

A young couple turns this once dreaded experience into a learning occasion. Wheeling their daughter in a cart through the aisles, they put into her hands some of the items they intend to buy, saying "rough" (a hemp doormat), "smooth" (a polyester scarf), "cold" (a pint of ice cream), "tickly" (the fuzzy ends of green onions), "pointy" (the tip of a pencil). The little girl giggles delightedly. Her tactile curiosity well satisfied, she doesn't reach and demand.

Here are other ways to enrich a baby's sensory experiences. Keep in mind that the child enjoys temperature, motion and rocking similar to what it knew in the womb.

The most important piece of furniture an infant can have is a rocker, with arms.

Children adore squeezing wet gushy substances through their fingers. Have a mud hole in the backyard where children can wallow. Those who play with their own feces are often those who have not been allowed to play in mud and sand, wet goosh. I cringe when parents say, "Go play but don't get dirty." Let's change that to "Go play and have fun getting dirty."

If a child screams during a repeated act of caretaking, look for a cause stemming from sensory discomfort. A mother told me her four-month-old son became hysterical every time she diapered him. I asked her to show me what she did. She removed the diaper, wet a washcloth with *cold* water and wiped the infant's warm crotch. I too would scream if a cold, sopping-wet washcloth were suddenly put on my warm crotch several times a day.

Hold a baby on your lap or in your arms during examinations by a doctor. Tell your pediatrician, if he doesn't already know it, that babies who are held cry less than babies placed on examining tables.

The more that babies' brain circuits are pleasantly stimulated, the more they feel loved. Is there danger of sensory overload? Can we harm a child by stroking and petting it too much? Yes, that's a possibility. As Beverly Hills psychoanalyst Maurice Walsh points out, "Neurotic parents may, by constant fondling, attempt to conceal their hostility to their child"—and thus create a seductive child. "There is reason to believe," Walsh says, that "much too much caressing may be as disturbing for personality development as too little."

Given our present social attitudes toward touch, there is less likelihood of a child being fondled too much than of it not being fondled enough. Infants can absorb tender caresses on their faces, hands, chest, abdomens, arms, legs, necks every time they are awake during their first months.

There is some indication that Nature takes care of the possibility of sensory overload. According to findings of Sir John Eccles, 1963 Nobel Prize winner for his work on the transmission of impulses between neurons, and his wife, Helena Taborikova, a neurophysiologist; "A very large number of nerve cells at the higher level of the brain do not excite other nerve cells to fire, but inhibit them from doing so. This inhibition prevents the brain circuits from becoming overloaded."[19]

How loved Deren Dusay of San Francisco must feel! He was born on December 12, 1978, to Katherine and Jack

Dusay, a psychiatrist and past president of the International Transactional Analysis Association. Katherine tells about her son's birth:

> We had a "home" delivery in a hospital setting. I gave birth on a kingsized bed, with my arm around Jack. The lights were dimmed so Deren's eyes wouldn't be offended. We had Sibelius playing in the background and we talked in whispers—no harsh voices for Deren's ears. When he arrived he wasn't spanked upside down. Instead he was tenderly placed on my tummy which has the exact dimensions and shape of his former home. Then he was moved to my breast where he eagerly suckled. After we were certain that his respiration was perfect (in about 20 minutes) Jack cut his umbilical cord and the placenta was delivered. We then shared our wedding wine and I put a drop on Deren's lips. He was completely alert and bonded with Jack and me in the most intimate way. The state-required eyedrops and shots were delayed at our request. The three of us stayed in the huge bed for two days. Wonderful! He never left us!

A photograph of mother, father and child snuggled in bed together, taken two hours after Deren's birth, is a paean of joy—a modern nativity scene.

3

Touching in Family Life

I was a free-lance reporter for the *New York Times*, and I was interviewing Marilyn Monroe. Marilyn sat at her dressing table, toying with a powder puff in her hand. I asked, "Did you ever feel loved by any of the foster families with whom you lived?"

"Once," she replied, "when I was about seven or eight. The woman I was living with was putting on makeup, and I was watching her. She was in a happy mood, so she reached over and patted my cheeks with her rouge puff. For that moment I felt loved by her."

Reenacting the experience, Marilyn dabbed the powder puff on her cheeks. The memory brought tears to her eyes and she was silent, thinking back to that rare moment of childhood joy. It brought tears of compassion to my eyes to see the famed sex goddess weep as she recalled how meaningful that playful touch, so casually bestowed, had been to a little girl starved for love and affection.

There are many Marilyns among us, people who grow up rarely knowing that quick response of joy, that lift of gladness, that we get when we are touched in a manner that says "You are cherished." Mothers and fathers may prepare food for us, they may chauffeur us around, they may give us money to buy an ice-cream cone or go to a movie, but nothing they do registers as deeply on us as do their squeezes, pats, strokes and embraces. It is not what our families give to us or do for us that makes us feel their love, it is our bodily sensations when they touch us. Tactile sensations become our emotions. We can receive no greater assurance of our worth and our lovability than to be affectionately touched and held in the cradle of family life. Knowing that we are valued sends us into the world with some magical inner strength to deflect life's slings and arrows.

Psychiatrist Robert Coles tells of an experience which made him aware of the power of touch to give strength during extraordinary stress.

> In the spring of 1961 I was spending my days with black southern children struggling to desegregate schools against extremely discouraging odds. In the case of four New Orleans girls, each six years old, the resistance was fierce—a source, eventually, of national shame and sorrow. Every day, mobs heckled those little girls mercilessly. Death would be their fate, they were told.
>
> Yet these children, all from impoverished, vulnerable backgrounds, did not once yield to their tormentors. Eerily, their school was empty of other pupils; all the white boys and girls had been withdrawn by their frightened, confused or angry parents. But these children went to their classes regularly, studied hard, and did fine at home. They survived, learned, and grew up solidly and well.
>
> Their teachers, not to mention me, a child psychiatrist anxious to see how young people handle stress, were full of wonderment because many badly harassed schoolchildren would prove themselves remarkably resilient.

Later, a Louisiana black mother told Coles what she believed had given her daughter strength to withstand the pressure:

> My child comes home from school, and she's heard those white people shouting, and she's not going to show them she's scared, not for a second, but she is scared. I know she is. And the first thing she does is come to me, and I hold her. Then she go gets her snack, and then she's back, touching me. I'll be upset myself, so thank God *my* mother is still with us, because I go to her, and she'll put her hand on my arm, and I'm all settled down again, and then I can put my hand on my daughter's arm. When my mother puts her hands on me, and I put my hands on my child, it's God giving us strength.[1]

Nothing binds up psychic wounds like the bandage of a hug—a warm clasp around a hurting human being. When I am unusually irritable, flaring up over incidents I would normally take in my stride, snapping in an edgy voice on the telephone, feeling close to tears, I ask myself, "What's wrong? Why are you behaving this way?" And I become aware that I need a "fix." I haven't been held close for a while. My soul needs soothing; loving contact is the only thing that can do it.

These were exactly the symptoms I was feeling while writing this chapter. I had not seen my male companion for a week because of our busy schedules. I felt my energy drooping, my enthusiasm for my work oozing away. Then, unexpectedly, he stopped by in the middle of the day. We sat on the sofa and I stretched out, putting my head on his chest while he cradled me in his arms and we talked, holding hands. That embrace reenergized me, my delight in my work restored.

John Schneider, a Lansing, Michigan, newspaperman, says he goes around with a "small hollow spot" on the days when he hasn't held his family. "There is a minor itching in my soul, a subtle tug of expectation that stays with me until my wife and kids and I squeeze it away together."[2]

We Americans may be prodigal in material goods but we are stingy with little gestures that can bring happiness to those we love. Observing parents and children on beaches, in parks, in playgrounds and in shopping malls, researchers recorded that children initiated and sought affectionate and comforting touch contact to which parents often did not respond. Typically, a small child would touch her mother's cheek and put her arms around her mother's neck but the mother did not reciprocate. When a child hurt himself, a parent was apt to comfort him by offering crackers, candy or cookies rather than physical comfort. (This is undoubtedly the beginning of obesity for some who, as adults, reach for food for the stomach as a substitute for what they would prefer—food for the skin.) Parents touched children mostly for purposes of cleaning, such as wiping a running nose; of controlling their behavior, so they wouldn't wander too far; of punishing them for taking another child's toy. A father, "watching" over his child in the playground while reading a newspaper, did not once make physical contact during the hour I observed him.[3]

No parent grabbed a child and gave it a joyful hug, cheek rub or back rub, or made any other physical gesture that would express pleasure in being a parent and having a child.

The question of how much pleasure we truly get from parenting is raised by the observations of social scientist Paul Rosenblatt of the University of Minnesota, who recorded behavior in public places of couples with and without children. Childless couples touched more frequently, walking arm in arm, holding hands, kissing and making other affectionate gestures. Couples with children touched, talked and smiled much less. Are we, as myth has it, a child-centered culture? I see us spending a lot more time shopping and buying for children than we do giving them one-to-one contact and caring. We are actually a possession-centered culture; we devote ourselves to what our children possess rather than to the children themselves.

If we asked parents, "How come you don't touch and hold your child?" some would justify their indifference by saying, "Oh, my child hates being touched; she runs whenever I try to grab her to hug her." *Grab* is the key word. We have all seen parents or other adults forcibly restrain a squirming child with hugs when he wants to be off exploring; squeeze a child so tightly that he yells in protest; pounce on a child absorbed in play to embrace him; force a reluctant child to hug and kiss friends, relatives and even strangers in hellos and goodbyes. John Holt calls this behavior the "cuteness syndrome"; we force kisses and caresses on children because we find them so adorable and cute.

Children are not born disliking to be hugged. Such a response is conditioned into them by grownups who hold too tightly or painfully, at the wrong times and in an annoying manner. To restore children's inborn delight in being touched, parents might post this reminder on a child's bedroom door: "Give us bouncing, not pouncing, tender touching, not tight touching"—which also serves as a notice to friends and relatives addicted to "cuteness."

A rationalization of some parents who don't offer physical affection is that "it takes time. Who's got time to hug or touch everybody in the family every day?" Tactile expression of love requires no extra time; it is easily bestowed while other activities are going on. As you walk through the house and pass a family member, it doesn't take extra time to draw an affectionate hand across that person's shoulder. It does not take any more time to cup a child's or spouse's cheek in your hand for an instant as all of you are scattering for work and school in the morning than it does to speak the four words of the goodbye ritual, "Have a nice day." As you place a dish of food on the table with one hand, it takes no extra time to press with your other hand the back of the neck of someone seated at the table. It says, "I'm glad you're here. I'm happy to be providing you with food." There are many sweet ways to show what we mean to one another, without taking extra time to do so. Such small gestures carry large meanings.

In a lengthy analysis of why the American family is breaking down, *Newsweek* cites television, taxes, inflation and poor family counseling. There is no mention of the effect that the deprivation of so basic a need as touch might be having on the disintegration of family life. Jack Panksepp of Bowling Green State University, an authority on the effect of brain chemicals in social bonding, reports in *Brain/Mind Bulletin* that satisfying interaction, including touching, causes the brain to release or synthesize chemicals that create a social "high," and that young people who miss out on such reward may turn to narcotics seeking that same high. He says addicts are giving clues to their feelings by personifying drugs, calling them by pet names and describing them as "my friend," "my family," "my loved one."[4]

Analyzing why some young people are drawn to cults, Arthur Janov says, "Cults and related movements offer a new family. They provide the follower with new people to worry about him, to offer him advice, to cry with him and, importantly, to *hold him and touch him.* Those are unbeatable attractions."[5]

My years of counseling and teaching families have taught me that it is unlikely that a child or spouse living in a tactile, loving environment would feel any need to take off through running away, using drugs or divorcing. When we are in psychic pain and there is a warm, caring, emphathetic person at home in whose embrace we can cuddle and talk of our anxieties and heartaches, or just cuddle and *not* talk, we feel better. Families who do this become closer. We do not run away from a milieu in which our need for touch and empathy strokes are well-filled; it is a search to fill such needs that is likely to make us want to leave.

Why don't American families take advantage of the "unbeatable attraction" of holding and touching? (When the wife of an American businessman who had lived in France returned to the United States, she was asked, "What was the single greatest difference you experienced in French life?" She answered, "The physical affection French families give each other.")

Our discomfort is a reflection, as we have seen earlier, of the still-powerful Puritan heritage that taught us to equate touching with sexuality. In the past I have done that. I remember an incident when my daughter was about three or four and she asked to see "the hole I came out of." I opened my legs and pointed to the vaginal opening. What a miracle to a child, that she had emerged from such a small opening! She asked to touch it. I said no. As long as my daughter's curiosity involved the sense of *sight*, I was comfortable. But when it involved the sense of *touch*, I became uncomfortable. It was normal for a child to want to go from seeing to touching, a natural progression in which we seek to experience with more than one of our five senses. She was doing so from the stance of "I want to touch and know what it feels like," not "I want to be sexy with you." That's the automatic meaning that I, a grownup, gave to a child's innocent curiosity. Now I would handle this differently, saying matter-of-factly "yes" when she asked to touch "the hole I came out of." (There is a fine line between the possibility of eroticizing a child and of respectfully handling its natural curiosity about our bodies. Families find they must often deal with such a dilemma. They usually base their handling of the problem on their individual belief systems, as well as the pattern of how they answer a child's other, nonphysical, questions as well.)

Part of our Puritan heritage is the incest taboo, the greatest factor causing parents to feel uncomfortable about touching children, especially adolescent children. We're fearful, because we are capable of pleasant sexual sensations when we hug, caress or even just look at our children. A father writes to the "Dear Abby" column:

> I'm a middle-aged, happily married husband and the father of three wonderful daughters, 18, 17 and 14. For the last year I've found myself becoming embarrassingly stimulated by seeing my daughters in various stages of undress, lounging around in see-through nighties. Without revealing my problem to my wife, I suggested she impress upon the girls the need for modesty, even around their father. I never would have touched my daughters, but the sight of a beautiful, nude female can excite any normal male and he need not be perverted or a dirty old man to be aroused by his own daughter.[6]

Since just the sight of children can sexually stimulate a parent, we can easily understand how great the arousal might become if parents and children were to give each other a lot of affectionate touching. Our uneasiness creates distortions in the behavior of families. Living under the same roof, families go for years without ever making physical contact, except for special occasions like a wedding, when daughters might dance with fathers, sons with mothers, or brothers and sisters dance together—all of them doing so self-consciously.

A mother wants to express affection to an adolescent son. Secretly he would enjoy, as he did in childhood, being held close to his mother's bosom; instead he snarls, "Leave me alone." A father, departing on a trip, starts to kiss his daughter goodbye; he is hurt when she turns her mouth away and offers a cool cheek instead. Brothers and sisters, ashamed of a desire to be close, engage in mock wrestling matches or even physical fights, more socially acceptable than hugging as a way of being close.

Small children may be deliberately naughty for the payoff of being spanked or slapped. Sharp stinging touch strokes are better than no touch strokes; they at least provide needed internal stimulation. Young people become hypochondriacs so they can lie in bed and be physically cared for by parents who have an internal message that it's okay to cuddle a sick child but "not nice" to cuddle a healthy one. My mother's most loving gesture would come when we were sick. She used an old folk-medicine manner of checking our temperatures—putting her lips on our foreheads. I always enjoyed that soft brush of her lips there. I would have treasured that delicate feel of her when I was healthy, not sick. As an adult, I made up for it by frequently, spontaneously, kissing the foreheads and the tops of the heads of my own children. I still do that with people, of all ages, whom I care about.

The closeness most of us yearn for in family life is not sexual; it is not directed at eroticized parts. Families who are intimidated by the feel of one another's body bumps and bulges, breasts and genitals, would lose their inhibitions if they acknowledged, "Yes, occasionally we may have erotic feelings. They are a normal part of life. We have no intention of doing anything about them. So let's get on with feeling comfortable being close." In counseling I ask families how they feel about one another's bodies and whether their feelings keep them from physical closeness. They almost always say yes. I ask if they would like to stop fighting the battle of the bulges by embracing right then and there. They start laughing. Then, overcome by the unaccustomed intimacy, parents and children are often moved to tears.

Sensitive to the incest taboo, some parents worry about the propriety of an activity they would enjoy but fear might be arousing—cuddling in bed with a child while they all watch television, read, play cards, talk or occasionally even sleep together all night. I have no ambivalent feelings about such family closeness; I have a joyful memory of crawling into my parents' bed once and being allowed to spend the night. I can still feel the blissful warmth of their bodies, the thump of their heartbeats, and smell the slight gasoline odor of my father's skin. (He was in the dry-cleaning business.) This dilemma exists only in affluent societies. In poor countries families aren't troubled about sharing a bed. "Pretend," I tell families, "that you live in a poor country and all you can afford is one mattress."

Our inhibition makes us a nation of "pelvicphobes," relatives who greet each other A-frame style, clasping heads and chests, but from the waist down, standing with our bodies apart. Heaven forbid that we should get a mild tingle of pleasure from holding close a grown child who was created out of our own bodies, or a sibling or other relative through whose veins flows the same blood as ours! Some of us hug everybody, not just relatives, in this manner; we learned how in the bosom of family life.

I was amused to see columnist Erma Bombeck and her stepfather, who helped rear her from a young age, greet each other when he appeared on the *Good Morning, America* television show to celebrate Father's Day. Erma and Dad gave each other a "proper" All-American A-frame hug, the same way my male relatives and I do.

Other styles of hugging—the "Baby Burp," the "Drumbeat Tattoo" and the "Chimpanzee"—also reveal how uneasy we are with bodily closeness. Have you ever been hugged and felt that something was wrong but didn't know why? It felt as if the hugger didn't mean it but was just enacting a ritual? Chances are you felt this way because the hugger beat *rat-a-tat-tat* on your back with his fingers as if he were nervously or impatiently beating a drum. Or he patted rapidly as if you needed burping. Or alternately flapped his left and right hands, which feels like light slapping rather than a loving caress. That's how chimpanzees hug.

When a hugger gives me any of these embraces, his hand talk—the drumming or fast patting of fingers or the slapping of palms—says to me, "I'm not comfortable doing this. I'd like to get it over with quickly." I feel uneasy and wish I had not been embraced at all. But when a hugger presses his palms with fingers closed on my back or rubs his hands up and down caressing my back, this makes me feel good. It says, "I like you, I'm feeling good doing this. I hope you are too."

Check this out for yourself by making a hug test on a person you would enjoy embracing. Tell him you are testing something you read in this book. Then tell him you will give him different styles of hugs. Ask which he likes best. Exaggerate your finger and hand movements to be sure each hug is distinctly different from the others.

Give an A-frame hug, each clasping the upper body and sticking out the rear end.

Give a baby-burp hug, rapidly hitting his back as if you were trying to bring up gas.

Hug like a chimp, flapping hands from the wrist down, right hand, left hand, right hand, left hand.

Then give the full-bodied, full-palm hug with fingers closed, rubbing his back.

I demonstrated these hugs with the hostess on a Los Angeles television show. Many people phoned to thank me. Amused with their new insight, they said, "That's the first time I knew I was a drumbeat hugger" or "that I hugged like an ape." A minister made a poignant comment—"I'm large and muscular with long arms. My wife is tiny, five feet tall. Because I'm big and she's small, I'm always the hugg*er*, never the hugg*ee*. You made me aware I want to feel her arms around me, not always mine around her. I want to be babied by her. We did it and it felt great."

If Ralph Waldo Emerson had taken this hug test he probably would have said, as he did in one of his aphorisms, "I hate the giving of the hand unless the whole man accompanies it."

At times of high emotional intensity or family trauma, hugging behavior changes. The actual death or near-death of a relative routs our inhibitions. When kin gather for a funeral, or if a dear one miraculously escapes alive from a plane, train or car crash, we are no longer pelvicphobes. We fall into each other's arms, consoling or congratulating each other with full-bodied hugs. We want to get as close as we can to the whole body of those we care about while we are fortunate enough to be alive.

When families lose someone to illness, accident or war, they often express regrets, "Why weren't we closer when we had the chance?" Gene Mullen, an Iowa farmer whose son's death in Vietnam inspired the book and television movie *Friendly Fire*, regrets that the last time he saw his son, all he did was place a hand on the boy's shoulder and say, "Mikey, be careful." Columnist Mike Royko, lamenting the loss of dear ones in the 1979 O'Hare Airport crash, told of survivors' grief that "dear familiar hands would now be left unsqueezed" for eternity. At a memorial service dedicating a school library to the memory of a daughter who had died from a rare disease, a Los Angeles mother urged weeping families to be fearless in their displays of affection: "To all of you, young, old and in-between, love each other. Don't hesitate to hug and say I love you. There will come a time when that person won't be around anymore, and then you'll say I wish I'd done this or I wish I'd done that. Don't let that happen to you. If you feel like it, reach out."

Families seeking psychological help say their most frequent problem is "inadequate communication." We fail because we are using only part of the equipment Nature gave us to convey our meanings—speech. We need to use another component of communication—the language of our tissues. Our skin, our body warmth, our physical beings speak a rich vocabulary; with it we can make each other feel protected, appreciated and validated. The vocabulary of our tissues can tell us that we are forgiven; it can congratulate us; it makes us feel understood. I remember one time, when I was about eight years old, receiving an award in school, coming home and excitedly telling my mother. She said, "Congratulations." But I also needed closure for my accomplishment by being held. When she didn't do so, I felt bereft. The glory vanished.

At times we may be too embarrassed or too emotionally upset to put our feelings into words. "Tissue talk" can say it all for us. A mother and an adolescent son quarreled, and he ran out of the house in anger. The mother became increasingly agitated that he might run away. Hours later, much to her relief, he walked in, looking scared about the reception he would get. Wordlessly, she put her hand on his shoulder and gave it a squeeze. "It's all right," her touch was saying, "whatever separates us can never be as strong as the bond that holds us. I'm glad you're back."

A common experience, the daily homecoming, provides a good example of how our tissues can speak louder than our tongues. Many of us do not feel like talking the moment we come through the door. We may be exhausted from the pressures of the day. We need time to quiet and "center" ourselves from the sensory overload of the outside world. Family members misunderstand this need for quiet as indifference or lack of communication.

Touching is a fine alternative to silence or withdrawal. Tired people can acknowledge one another's presence by a greeting touch that speaks volumes—"Right now I'm overpeopled and talked out. I'll catch up with you when I'm rested." Most of us would be happy to get a squeeze of the hand, a kiss on the back of the neck, a touch around the waist, a tap on the arm. Wives who receive such homecoming strokes stop complaining that their husbands don't communicate. The husbands aren't *talking* more, they are *touching* more.

A wife said to me, "It's strange, I used to be upset that my husband wasn't talking to me as soon as he walked in. I realize now I don't even want conversation at that moment. I'm tired, too, and not feeling talkative. I want that touch gesture from him more than I want his conversation."

Another woman who is home with small children all day has, mostly with tissue talk, changed the emotional climate of her husband's arrival. "When he'd walk in and ask, 'What kind of a day did you have?' I'd rage at him about how the kids misbehaved, how everything went wrong, implying it was all his fault. I'd sulk all evening. Now, if it has been that kind of day, all I answer is, 'Just hold me,' and I fall into his arms. I don't have to say another word, he *knows* what kind of day it's been. After a little nuzzle, my rage is spent. Then I feel friendly to him."

I use tissue talk to give couples the feeling of what it is like to have someone "on your back." One of the couple bends over and the other lies on his or her back with arms around that spouse's shoulders. "Now," I instruct the spouse who is bent over, "walk through life with your husband [or the roles may be reversed] on your back." She struggles forward, dragging him. The two of them instantly get the feel of what's going on in their bodies when a spouse is on one's back, too dependent, demanding, smothering.

Pianist Manfred Clynes, fascinated by the meanings we convey through tissue talk, has conducted experiments of what he calls our "sentic cycles"—measuring the electrical impulses we generate through our fingertips when we express emotions like anger, hate, love, sexual passion, joy, reverence. We are able to "read" with great accuracy the fine subtleties of one another's meanings when we are touched.[7] (Even small children can learn to identify the different meanings of touches. The Fargo, North Dakota, Rape and Abuse Crisis Center; the University of Minnesota School of Medicine; and other groups concerned with child abuse and molestation, are teaching children to know the difference between a "red flag" touch—one that makes a demand upon a child—and a "green flag" touch—one that is genuinely loving and not manipulative.)

If yours is among the many families who don't touch, especially after children get older, here are some ways to start. At first you may get a response of "Hey, what's with you?" But don't get discouraged. People may be embarrassed and back off when they are the recipients of new behavior and don't know what to make of it. As with any change, it helps if we first *tell* others of our change. A useful beginning would be to discuss this chapter.

Tell your family that there are several styles of hugs that people typically give each other. Hook their interest by asking them to guess what those styles are. Demonstrate the hugs by embracing each member of your family. Ask if they feel the difference. What name would they give each style? Then tell them the names that I have given to hugs. Ask your family: "What kind of hugger have I been?" "What kind of hugger do you think you've been?" Have fun demonstrating; treat it like a game.

To see how well all of you know one another, find out if, blindfolded, you can recognize one another. Blindfold each member in turn, and then have the blindfolded one feel the hands and fingers of all the others. Remove rings and watches that could be a giveaway. Mix up the sequence of hands to make it harder. Are these Dad's or an older brother's fingers? Can the blindfolded one feel the difference between male and female fingers? If the blindfolded person guesses correctly, what clues did she (or he) use? Someone's big knuckles compared to another's small knuckles? Pointed fingernails compared to blunt nails? Who in the family has the shortest thumb and the longest pinky? A little girl described grandma's hands as feeling "rubbery."

Play the same game with toes and feet. Whose bare feet are these? Whose toes? Who is the most ticklish, the least ticklish, on the soles of the feet? Among my fond memories is playing "footsie" with young relatives. Two of us would lie on our backs on the floor or on a bed, put bare feet to bare feet and pump our legs back and forth, seeing how long we could keep it going until our feet lost contact, whether from fatigue or sweaty soles.

You can also play it with cheeks and lips. Can you tell just by the bone structure of cheeks whether it is a male or female face? Trace with your forefinger the lips you are trying to identify. What is the difference in the feel of a child's cheeks and lips, of a grownup's? Or do they feel the same to you?

Don't make these competitive games or give prizes for the most correct answers. These are loving games meant to open the path to greater family closeness, the awareness of one another's size, feel, texture of skin. Once a family learns to enjoy intimacy, you find many ways to make contact. Arriving at the home of my grandsons and finding them lying on the floor crayoning, I put my hands on their behinds and, moving their butts from side to side, sing out: "Rock-a-bye tushies, down on the floor." The two boys giggle happily at grandma's antics that feel so good.

Educator Sidney B. Simon, author of *Caring, Feeling, Touching*, recommends "back rubs" as a solution to some discipline problems. He says:

> One of the major domestic problems of this nation is getting kids to put dirty socks and underwear in the hamper. Parents shout, punish, cajole, but nothing seems to work. A lot of energy can be saved and more positive attitudes produced if the parents put dirty underwear in the hamper themselves but establish a rule that each time they do it, they are entitled to a five-minute back rub from the child who plays litter-bug. Littered laundry, instead of an exchange of shouts and recriminations, sets off an exchange of kindnesses. A back rub will build more character than any number of pieces of underwear picked up and put in the hamper.[8]

A Gallup Poll asks Americans what they want the most. "A happy family life" heads the list. To achieve that, we need to talk to one another as Nature meant us to—with our tongues and with our tissues. A favorite cartoon of mine shows a mother and two children telling Daddy, who is lying on the sofa and reading, "It may interest you to know touching, caressing—TACTILE displays of affection are what make elephant families so damn happy."[9]

Happy families can affect the world. Lao Tze, the Chinese poet, wrote two thousand years ago:

> If there is right in the soul, there will be beauty in the person
> If there is beauty in the person, there will be harmony in the home
> If there is harmony in the home, there will be order in the nation
> If there is order in the nation, there will be peace in the world.

4

Touching in Sex

Gestalt therapists sometimes use a technique in which body parts talk to each other. It is a quick and effective way to get to the core of problems that we tend to camouflage with depersonalized statements. Listen to this angry "conversation" between the genitals of a couple in sex therapy.

VAGINA I feel like screaming at you. I'd like to have knives around my lips and cut you off when you're in me pumping away.

PENIS What are you so bitchy about? Dammit, pump back. You're not soft and inviting, you're dry. You don't throb as if you're enjoying this. I need something soft and moist to lie down inside of. You frustrate me. I'm mad.

VAGINA I hurt when you get me up in the middle of the night and start plowing in me. You insensitive jerk. That hurts when I'm dry and you come pushing. Your body weighs a ton on me.

PENIS Yeah, and I'd like to crush you with my weight. Your body's like a stone statue, not giving. I want to feel arms around me pulling me closer. My butt wants warm hands pressing me deeper inside you.

VAGINA What kind of act of love is this? I hate you right this minute. Oooh, that hurt, hitting my walls with your hard, thick skin.

PENIS I'm coming, I'm finishing. There. O-o-oh, what relief. Please don't turn away yet. Just hold me for a minute. You sure aren't any help. Boy, am I going to pay you back for this!

VAGINA At last, it's over. I feel so sad. I want to cry. I need tenderness. I'm so hungry for slow, loving fingers all over me, as if someone cared about me.

This may sound far-fetched but it's realistic. If millions of male and female genitals were to have a dialogue, they would say some of these angry, painful, harsh things. The same plaintive cry runs through the words of many couples. The husband says, "My wife *never* makes an overture. If I didn't approach her, I think we'd go for years without sex. I'd like it if once in a while, while we're sitting watching television or reading, she'd reach over, unzip my pants, and fondle my penis. But she never makes a move to touch me."

The wife says, "My husband doesn't fondle me enough. He gets into sex too fast. I need more warming-up by being touched. I'm afraid to touch him because that automatically means sex to him. I want to be caressed and held before sex."

Underneath all these expressions of unhappiness is a plea for more gentle and prolonged tactile experience as part of the sex act. During the most intimate act we perform in our entire lives—making love—we deny each other the most gorgeous pleasure we will ever know—the light, soft, easy, slow motions of our hands and fingers. Masters and Johnson, the pioneer sex researchers, say: "The feeling that sensate pleasure at best represents indolence and at worst, sin, still permeates society sufficiently to influence the affectional sexual patterns of many relationships."[1]

After centuries in which pleasure was labeled as licentious and hedonistic, we are having a hard time adjusting to a new pro-pleasure ethic. We need to be given psychological and societal permissions to touch our bodies in ways designed to do nothing more than yield pleasure. Many couples, spending hours in bed in pursuit of pleasure, share uneasy, gnawing feelings that "we really ought to be getting up and doing something worthwhile." That having tactile pleasure *is* doing something worthwhile is a new idea that we have yet to become comfortable with.

New freedoms always take getting used to. Never before having had freedom of control over our own bodies, we have to be taught how. I too have had to learn this. During the question-and-answer period after a lecture someone once asked, "Do you think we ought to be teaching people what to do with their fingers and hands as part of sex?" The audience laughed. I answered, "No, I don't think we need to teach that. When people are psychologically free to enjoy sex, removed of the burden of sex primarily for procreation, they'll know what to do with their fingers and hands."

It's fifteen years later now, and I am fifteen years wiser. Yes, we certainly do need to teach people what to do with their fingers and hands. In fact, every theory of treating sexual problems is based on teaching us what to do with fingers and hands. The major technique used by all sex therapists is to teach couples how to touch.

Masters and Johnson originated touch exercises known as Sensate Focus. In a two-week program couples are instructed to start out touching each other's body lovingly, slowly, tenderly, gently, playfully, everywhere but in the genital and breast areas, in the privacy of their rooms. They take turns, one being the "giver," the other the "getter." They are told that it is okay to receive erotic pleasure for themselves without feeling any need to reciprocate during that touch session.

Touch assignments show couples that when they feel exquisite pleasure and sensual stimulation without the burden of having to perform the sex act, they can repeatedly experience arousal and desire. Over and over, an impotent male sees evidence of how, when the pressure to achieve penetration is removed, he is capable of achieving and maintaining erection. The nonorgasmic female learns how her body can respond with intense desire for orgasm when her face, lips, eyes, ears, abdomen, knees, shoulders and buttocks are stroked, and her clitoris, vagina and breasts, her most sexually sensitive areas, are deliberately left untouched. From such touch, paroxysms of joy and desire can surge through the body until it feels, in the words of one couple, "like a huge penis or vagina crying out for genital contact," throbbing with the need for release through orgasm. Eventually in the two-week program, couples are given permission to engage in intercourse, having first learned how to touch each other.

Another noted team of researchers, William Hartman and Marilyn Fithian of the Center for Marital and Sexual Studies in Long Beach, California, has designed other touch techniques:

The Foot Caress, in which couples bathe each other's feet and legs. Try this and you'll become aware of how soothing it is to have loving hands fuss over you with soap and water. Sit on a chair with your feet in a pan of warm water while your partner sits on the floor in front of you. Close your eyes and experience the water being poured over your calves, ankles and feet, the creamy soap suds massaged up and down your legs. The part of this exercise I especially like is having my toes tugged at and then exercised in a circular motion. I become aware of toes as separate parts of myself.

The Hair Caress, in which couples gently wash, comb or brush each other's hair. It can be spine-tingling to feel the warm breeze from your lover's breath tickling through your hair.

Breathing Caress, in which a couple lie closely, face to face or spoon fashion, one's back against the other's front, and synchronize their breathing. It becomes hypnotic to feel each other's rhythmic abdomens as you inhale and exhale.

The Body Imagery Exercise, in which you stand nude in front of a three-way, full-length mirror and touch every part of yourself from the top of your head down to the soles of your feet, not skipping the belly-button or pubic hair. You say out loud what you like or dislike about that part of your body. After voicing these feelings, the small size of one's penis or of one's breasts may never again seem so important.

When I did this exercise in a workshop, I was surprised to hear myself saying, while touching my shoulders, "This is what I like best—the looks and size of my shoulders." I had never even thought that before; the statement came out spontaneously. A one-breasted woman proudly told how she loved her mastectomy scar because it meant she was still alive; a man loved the huge gash down his middle because it meant he had had successful heart surgery. That moving experience forever changed my distaste for surgical scars; what seemed ugly became beautiful.

Most innovative of Hartman and Fithian's touch exercises is the Sexological Examination in which couples are taught to lightly finger each other's genitals, pelvises and anuses to determine their spots of greatest sexual responsiveness. It's well-known that the most sensitive area in the male is usually the *corona*, or "crown," of the penis; the shaft and base may have lesser sensitivity. The woman strokes the different parts while the man tells her which part feels high-, medium- or low-intensity arousal. The man is taught to put a well-lubricated, gentle finger inside the vagina and to move it clockwise from the top down the right side to the bottom and then back up the left side, as if the inside of the vagina were the face of a clock. He helps his partner identify her "time zones" of greatest pleasure. Being touched at "two" and "four o'clock" may not do a thing for her, while being touched at "six" and "eight o'clock" may transport her into ecstasy.

An aspect of sexual touch that often leads to misunderstandings is the timing of it. A sudden touch makes us undergo the startle reflex, a reflex we are born with. Our startled bodies may perceive such a touch not as enjoyable but as threatening. Your lover puts his moist tongue into your ear and you reflexively jump. He mistakenly thinks that you don't enjoy this sex play. You do enjoy it—but only *after* you've reached a state of arousal. You and your lover start foreplay when he suddenly starts going down on you for oral sex. But you need more fondling first, so you hold him back by pressing your hand on his shoulder. He misinterprets your touch message as meaning that you don't want oral sex. You do, but *not yet.* As I mentioned earlier, some men make sexual overtures that are just the opposite of what our sense receptors crave. They give an "affectionate" smack to a woman's bottom or they grab her breasts from behind. (I call these acts "copulation clutches.") He doesn't understand why he gets an angry response. When I have said this to audiences, women told me, "You've just explained why I get angry instead of amorous when my husband approaches me. He startles me and I get mad."

The pace and pressure of our fingers is also important to good sex. Early in the act we enjoy a "feather message," long, slow strokes with airy pressure. Only after we are aroused do we want short, rapid, intense strokes with heavier pressure. A partner whose finger probes prematurely with short, hard motions can literally turn off pleasurable sensations rising in us.

Our receptors also make us sensitive to temperature and texture. A woman says, "It's cold in here," and the macho lover replies, "Don't worry, baby, I'll make you hot." He doesn't know that the chill can keep "baby" from becoming sexually hot, because our bodies contract when we are cold and they secrete sexual fluids more easily when we are warm.

If the skin on a man's hands is rough and coarse, it can feel like sandpaper scraping his partner's body.

Two items easily solve temperature and texture discomfort: a moistening lotion whose lubricity and fragrance you like, and a large cotton sheet blanket. Creaming and caressing each other's hands can be highly pleasurable, because our palms are so erogenous. (When I was a teen-ager, when a boy circled a girl's palm with his finger, he was saying, "How about a screw?" It was considered insulting—deliciously insulting—to have a boy do that to you.) A cotton blanket makes a cozy tent to throw over a lovemaking twosome in a cool room. Its lighter weight is more comfortable than a heavy blanket, and its texture is not itchy on bare skin as wool or polyester can be.

A little-known aspect of sex touch is that we have touch "sidedness," related to the functioning of our left and right brain hemispheres. Physiologist Barbara Brown says, "Skin talk of the right side of the body is louder and clearer than that of the left, regardless of whether we are right- or left-handed."[2] Couples who lie in bed in such a manner that the left side of one is the same side always accessible to caressing might occasionally alternate positions so that the sensitive right sides of both can be available for touching.

Strangely, the most used furniture for sex—bed—sometimes may not be the best place for it. Beds don't have built-in armrests. I prefer cuddling on a sofa with arms that can support my lover's arm when I'm lying with my head on his chest. An elderly bachelor told me that throughout his long sexually active life he always preferred to make love on a sofa, because the back provided a base from which he could thrust harder. "In bed," he explained, "your body is moving in empty space; on a sofa you have its back to give you leverage."

Not only the furniture on which sex takes place, but the location as well, can be stimulating. Some lovers like the excitement of being in a semipublic situation. It's like playing "cops and robbers," to see how much they can get away with in public. In counseling, men have said that they like to have their genitals fondled by a spouse or lover while driving a car, enjoying the sensation of the penis hardening in their clothes. They get a special erotic thrill risking being seen by people in other cars. One client worked on a farm as a young man. Even though the whole family slept in one large room, he maintained a sex relationship with one of their daughters. During the night he and the girl would surreptitiously crawl into each other's beds and engage in sex, making a game out of how quietly they could do it and not awaken the others. As a result of such "cops and robbers" lovemaking—"Will I get caught, or won't I?"—he functioned best thereafter when he and his understanding wife began their sexual touching in public.

This phenomenon is far more common than many of us will admit. It happens because we may have had our earliest erotic experiences in semipublic situations, in cars in lovers' lanes, under the boardwalk at beaches, behind bushes in parks, in hallways of our homes, or as young boys masturbating with friends in a companionable "circle jerk."

My lover and I often sit on the same side of booths in restaurants (waiters are startled when we do that) so we can touch thighs while dining or I can rest my head on his shoulder while waiting for food to arrive. If we're feeling especially playful we might touch each other's genitals under the table. This play doesn't necessarily mean we're feeling sexy—it is just one more way to be affectionate and to share a secret nobody else knows, that under the table we're being "intimate" in public.

An exciting part of semipublic sexual touching is that it happens when we're clothed. Being touched through clothes can sometimes titillate more than being touched in the customary way and place, naked in bed. Such a repetitive start to sex becomes downright boring. Being lightly and delicately touched through our clothes introduces a variety of stimuli that our physiology craves. Clothing can provide pleasurable friction. A man who had difficulty maintaining an erection when naked found that when he wore his undershorts while engaging in sex, the sleek feel of the silky pima cotton helped to stimulate him.

In sensuality workshops participants learn to touch not only with fingers and hands but with fabrics. Two fabrics especially effective in arousing us to shivers and goosebumps of delight are chiffon—synthetic fabrics such as polyester don't have quite the same effect—and feathers. Next time you are with your lover surprise her by having a scarf and a feather handy. See what joyful hackles you can raise by lightly trailing either one down your lover's spine and around her genitals.

Foods also introduce new sexual touch sensations. Remember the scene in a Marcello Mastroianni movie in which he gallantly squeezes out cocktail spread from a tube on to his mistress's abdomen and then licks it off? Or the porno movie in which the heroine inserts grapes into her vagina to be sucked out one by one and eaten with gusto by her lover?

As we crusade to find the Holy Grail of happy sexual functioning we repeatedly find evidence that the way to win that loving cup is through sensitive touching. With clothes, without clothes, in private, in public, it is more the pressure, pace, temperature and vibrations of our touch, and less the shape, size, style or gender of our genitals that creates the ecstasy we seek. Ashley Montagu says ". . . it is highly probable that the frenetic preoccupation with sex that characterizes Western culture is in many cases not the expression of a sexual interest but rather a search for the satisfaction of the need for (touch) contact."[3]

Promiscuous men and women, women who work as prostitutes, and women who repeatedly have unwanted pregnancies have told researchers that their sexual activity was merely a way of satisfying yearnings to be held. Physician Marc H. Hollender, in the *Archives of General Psychiatry* reported on a study of twenty women who had had three or more unwanted pregnancies; eight said they were "consciously aware that sexual activity was a price to be paid for being cuddled and held." Touching before intercourse was more pleasurable than intercourse itself, "which was merely something to be tolerated."[4]

I believe that our deep need for touching makes us capable of engaging in single acts of homosexuality, though not necessarily of a permanent homosexual preference. Given the right circumstances, such as long absences from the opposite gender—in military service, scientific explorations or prison—many of us would find ourselves having sex with individuals of the same gender, less from sexual need than from sensual need. We can give ourselves orgasms through masturbation, but, deprived for a long time of bodily contact with another, we would engage in a situational homosexual contact rather than do without. The veteran whose life was reported in the PBS television documentary *Frank, A Vietnam Veteran,* told of "loneliness so savage it sent him into the arms of another man when he wanted to be held."[5] Maggie Kuhn, founder of the Gray Panthers, tells audiences that some older people, deprived of partners of the opposite sex, are turning to partners of the same sex so that they do not have to live out their lives with unmet longings for closeness to a loved one.

Regardless of our age, we all have the desire at some time to feel babied and nurtured by another, a pull to return to the conditions of infancy. Even if it is only for a few moments, we need to recapture the sensations of being taken care of and protected by someone else, of not needing to do for ourselves. For those few moments when we are encircled by another's arms we can stop being our own doers and providers. We enjoy a respite from the intense pressures of living. We enter into a higher realm of consciousness—are lifted above the ordinary plane of living. The feel of an embrace can do that for us.

The most exciting and breathtaking sexual response I have

ever experienced, the one that stays in my mind as no orgasm ever has, doesn't even involve genitals. It involves touch.

It was a scene in that classic French film *A Man and A Woman.* A race-car driver and the woman with whom he is falling in love, the mother of his child's schoolmate, are in a restaurant. Their hands are resting on the back of a child's chair between them. As their excitement about each other mounts, their fingers slowly, gradually slide across the back of the chair toward each other. The close-ups of the yearning fingers are interspersed with scenes of conversation and eating.

Sitting in the audience I became almost breathless with sexual excitement. I felt weak and dizzy and moist in the groin as I watched those fingers get closer, exactly the reaction the filmmaker intended. At last those fingers *meet*, and the two people touch each other for the first time. The whole screen seemed to explode with sexual electricity, and you could hear the audience gasping with relief as this powerful sexual scene develops.

Every one of us has sexual perception as unique as our fingerprints. No two people have sense receptors that respond exactly the same way to touch. This is why a man who thinks he is a great lover and becomes annoyed when a partner tries to tell him how she likes to be touched may actually be a poor lover. Lovemaking is best when two people teach each other, with their touch, what their areas of greatest pleasure are. Our most effective "sex organs" are not our genitals but our fingertips.

5

Touching Ourselves

In your mind's eye, get a mental image of what you do with your body when you are upset. What are your "stress gestures"? Do you play with your chin? Pull at an ear lobe? Stroke your beard or tug at your mustache? Chew on a fingernail? Do you rub fingers back and forth across a necklace, push and pull a bracelet up and down your wrist and forearm, twirl a button on your jacket, pick lint from your clothes, wipe and wipe and wipe your eyeglasses, pat the back of your head, rub your forehead?

These are typical tensional outlets; we are often not aware that we are doing them. No matter how calm we may think our outward appearance is, we show our inner turmoil by such self-touches. They may be saying, "I'm insecure. I'm uncomfortable. I'm feeling inadequate. I'm irritated. I'm bored. I'm impatient. I'm angry." I wince now, as I remember an early occasion when, as a public speaker, I spent an evening nervously pushing up my eyeglasses—to the exhaustion and irritation, I am sure, of the audience.

Our self-touches are of two types—voluntary and involuntary. We choose whether to engage in a voluntary touch, such as applying makeup, smoothing creams and lotions on to our bodies, adjusting our clothes, relaxing parts of our bodies by self-massages, and so on. At this instant I am resting my head on my left hand while my bent left elbow is resting on my desk, my left hand is massaging the back of my neck, and my right hand is writing this sentence. By holding myself this way, I am supporting tired neck muscles as I bend over my work.

But our self-touching is mostly involuntary; it is a reflex action arising from our autonomic nervous systems. You are on a trip, let's say, and you arrive at a hotel and unpack your suitcase. Suddenly you realize that you forgot that special dress that you wanted to wear. Involuntarily, you give yourself a slap on the cheek, "Oh, dammit. I forgot that dress!" With your self-slap you are saying, "I'm so mad at myself. How could I have done such a stupid thing?" You are punishing yourself for being careless.

Or you are walking along the street and you see a youth on a skateboard come sailing down a driveway and almost into the path of an oncoming car. Involuntarily, you clap a hand to your mouth. You have two feelings. One, you would like to scream out a warning. Two, you have undergone a lifetime of conditioning that makes you reluctant to create a spectacle by screaming in public. And so you gasp out a warning and clap your hand to your mouth, a gesture that expresses your shock and yet stifles a loud scream.

If we were to monitor every self-touch, we would find that our need to relieve stress triggers by far the greatest number of our touches. We touch ourselves many times a day as we displace our anxiety and tension from internal to external expression. When you massage your neck, shoulders or lower back, are you *really* saying that a person or a situation at that moment is a "heavy load," "a pain in the neck," or "a pain in the ass"? Do you rub your forehead to wipe away the "headache" you feel from someone or something, a responsibility or a burden you don't want? Do you keep pulling off and putting on your wedding band? Are you really saying that you are ambivalent about getting rid of or holding on to your marriage?

Our stress gestures are actually healthy outlets. Nature wisely forces us to engage in motions that help work off the adrenaline coursing through our blood. Adrenaline prepares us for "fight or flight," a gross motor activity. Most of the time we cannot fight or take flight—we may be behind the wheel of a car, sitting in a business conference, standing in a slow check-out line at the market. Nature helps us to get rid of some of our pent-up tension with the small motor activity of touching parts of ourselves. Desmond Morris calls these "displacement" gestures.[1]

But while we are relieving our own tensions with repeated motions, we could be creating tensions in someone else! When I see a person who, on a TV talk show, repeatedly tosses her long hair away from her face and, before the gesture is completed, tosses it away once again, it triggers peculiar dissonance in me. I get impatient, believing that it is wasteful for anyone to engage in nonproductive acts over and over; when that happens, it is sensible to change one's behavior. I wish that the woman with such a hairstyle would wear barrettes or bows, just let her hair dangle, or get a haircut.

A personnel manager observes the displacement habits of jobseekers and will not hire an otherwise qualified person if she believes his nervous gestures will cause tensions in coworkers and customers. A department manager in her company has written, in an evaluation of an employee, "I don't know why—he's competent enough—but I'm uncomfortable around him." "*I* know why," she said; "the employee's constant mannerisms of adjusting his tie, hitching up his pants, rubbing his finger up and down the side of his nose. You feel like grabbing his hands and yelling, 'Stop.' "

To help people see how they come across to others, our society has spawned a new industry of "Image Makers,"—

those who run workshops in the course of which they videotape people at such activities as sales presentations and job interviews, while conducting conferences or chatting at business luncheons or at cocktail parties. The purpose of the taping is to make an executive aware of stress gestures and other mannerisms, and to learn how he or she is unknowingly making others nervous with them.

They teach participants to "displace their displacements." If, for example, you nervously poke at a cheek, or scratch a knee, or press your fingertips together—all visible gestures—they teach you to displace your visible self-touches with invisible ones. Instead of pressing your fingertips together to form an arch, you learn to keep your hands quietly on your lap or at your sides but press your thumb and forefinger hard together, so nobody else is aware of what you are doing. A public speaker controls her visible gestures on the platform, but inside her shoes she is rapidly curling and uncurling her toes, working off tensions. When I speak I like to wear a skirt or trousers with pockets so I can dig my hands into my pockets and press hard against my thighs. It comforts me to be able to discharge tension with this hard pressure and yet not have it show.

Our major reason for touching ourselves is to deal with stress. We touch ourselves for other purposes as well:

Cleansing
Expression of joy
Healing
Beautification
Sexual pleasure

Cleansing. Every time you bathe or shower you stimulate your nervous system as you stroke your body, soaping it and washing off. This is the only form of acceptable self-touch all over our bodies that most of us allow ourselves. We are inhibited about stroking ourselves "on dry land" as we do when we are bathing. Our conditioning says, "It's okay to be intimate with our own bodies if we are doing it for the noble purpose of cleanliness but not for the base purpose of pleasure." We are willing to insert a finger into our body orifices as part of cleansing but we resist doing so just for pleasure.

Unrefined though it may seem, we even find pleasure in cleaning out debris from our noses. Putting a finger inside a nostril can be comforting and may feel good, but it is acceptable to do so only in private. How confusing this can be for a small child. He puts a finger up his nose because it feels good, but we tell him, "Stop that! That's not nice!" But then wrap a tissue or a handkerchief on the finger and put it in the nostril. This self-touch now becomes socially acceptable.

Once, at a resort hotel, I glanced across an open courtyard and saw a woman resting in a room opposite mine. She was reading and contentedly picking her nose, but she seemed so vulnerable that I turned away and closed the drapes. I did not want to intrude on her private cleansing act, which I was sure must have felt good to her, as it does to me when I do the same.

There is something likable about people's vulnerability. When I counsel couples I ask, "What is your wife (or husband or lover) doing at those moments when you feel tenderness toward her rise up in you?" I am no longer surprised at how often people say, "When I look at her sleeping face," or "When I see him sitting on the toilet seat." There is something childlike about all of us at moments when we expose our common humanness. That's how I felt about the woman picking her nose.

Our social attitudes toward nose-picking may yet change. Matt Thomas, a teacher of self-defense, goes to colleges to teach women how to ward off would-be attackers or rapists. He advises them that in some situations they should pick their noses. In the *Los Angeles Times*, a Stanford University co-ed reported that it worked. While waiting alone at a deserted bus stop at night she was approached by two men. Quickly she began picking her nose. The pair looked disgusted and left.

I decided to try this. As I was driving home alone late one night, a car drew up next to mine at a stoplight on a lonely street. The driver looked over flirtatiously at me. Instantly, I stuck my finger up my nose. The driver, probably revolted, roared off in disgust, and I doubled over with laughter. Now when I leave for a late night out alone at a meeting, my lover may say playfully: "Remember, if any guys approach you—" and he sticks his finger up his nose as a loving safety reminder. I am taken with this line of poetry by Dylan Thomas's widow, Caitlin: "A lot of warm vulgarity is incomparably preferable to a little bit of pinched niceness."[2] Paraphrasing this, I tell women, "Better some protective vulgarity than dangerous refinement."

Expression of joy. You may remember a classic World War II photo that showed a small boy in a foreign country ecstatically hugging himself because he had been given a pair of new shoes. Like this child, you may throw your arms around yourself when you are feeling euphoric. Or you may spontaneously clasp both hands, one on top of the other, in the center of your chest, your gesture saying, "I can't believe this wonderful, marvelous thing is happening to me." Men consider this gesture of joy to be "feminine"; they are more apt to punch one hand into the other or clap their hands against both sides of their heads.

I am a folk-dance enthusiast. Sometimes when I finish a dance such as a polka, whose lively pace I especially enjoy, I break into childlike clapping. I can't help myself. What am I applauding? Not the record that just played or the other dancers. I'm saying, "Wow! Yippee! That was great. What joy I'm feeling!"

Healing. Interestingly, the gesture of hugging ourselves as an expression of joy is also used to heal ourselves when we are sad or depressed. When a client was overwhelmed with despair, she would wrap herself in her own arms and rock back and forth. I asked her what the rocking and self-hugging were saying. "I'm all I've got, I'm all I've got," she repeated. We discovered that she, as a small child, had done exactly that when her mother abandoned her at an orphanage. She was trying to comfort herself and heal her grief. Nature came to her rescue, making her engage in self-touches and rocking motions to stimulate the secretion of brain chemicals that could make her feel better.

That's psychological healing. We are also constantly using self-touch for physical healing. We bang an elbow or stub a toe. Instinctively we grab the injured part, rub it and press on it to relieve pain. Napoleon, in his familiar gesture of keeping a hand inside his jacket, was actually massaging himself to ease the pain of a stomach ulcer.[3]

In the past decade or so in the United States, since the greater acceptance of Oriental techniques of acupuncture, we have seen the growth of self-healing techniques through self-touch. In 1971 James Reston, a *New York Times* editor, had an emergency appendectomy while touring China. Impressed with the effectiveness of acupuncture in relieving his postoperative pains, he wrote a series of articles that set off an avalanche of interest in the Western world in acupuncture, done with needles, and acupressure, done with fingers. Now we have acupressure, do-in, jin jin shitsu, and *Touch for Health*, all forms of touch-healing, and, more recently, reflexology, in which touch-pressure is applied to zones on the soles of the feet. Some of these, *Touch for Health* and reflexology, may require some touching by another person, but all these methods are being taught in the holistic-health movement as self-healing techniques.

Touch for Health, which combines the Western theory of physical manipulation with the Eastern theory of acupressure channels, was created by John Thie, a Pasadena chiropractor.

It was inspired by George Goodheart, a Detroit chiropractor, who discovered that when energy is blocked in our major acupressure channels, certain muscles are weakened. By massaging appropriate spots we unblock the energy, and the strength of our muscles is restored. We do need another person to help us test our muscles and, ideally, to do the follow-up massage of points on our bodies, but we may also massage these points ourselves and feel positive results. As part of preventive medicine, practitioner-instructors in twenty-two countries are teaching lay people to strengthen weak muscles, help alleviate pain and reduce physical and mental tension, by balancing the "*chi*" energy that the Chinese believe flows through every living thing.[4]

I once volunteered for a demonstration by Dr. Thie on a television program. He suggested I think of a sad experience in my life. I thought of the day my daughter moved away from home to enter college. Dr. Thie asked me to hold my right arm straight in front of me with my thumb pointed toward the floor, and to resist his pressure on my arm. He pressed. I tried resisting. I had no strength in my arm; it went down instantly. Dr. Thie massaged an acupressure point between my ribs on the left side of my chest. Then he lightly held his fingertips against my forehead between the eyebrows and the hairline for about a minute. Once again I held my arm out in the same position. He pressed. But this time, miraculously, my muscles were strong, and I was able to resist; my arm did not go down, and my sadness was gone. Now, whenever I'm especially tense, I self-massage those two spots and I feel better. (That experience was a powerful lesson for me in how our minds affect our bodies.)

In the previous chapter, "Touching in Family Life," I mention activities to help families deal with embarrassment or discomfort about making tactile contact with one other. I recommend that families have fun trying Touch for Health, muscle testing and acupressure massaging to see what results they achieve. Touching for reasons of health makes contact acceptable!

These forms of healing are based on meridians, which connect parts of our bodies through nerve pathways like networks. We have hundreds of meridians. Along each meridian are acupuncture or acupressure points, corpuscles which surround the capillaries in the skin, the blood vessels and the organs. Massaging an acupressure point by rotating it with our fingertips, we help create bursts of electrical energy, which stimulate blood flow.

The fact that we do not yet know why certain meridian points are connected with certain other meridian points is an indication of how much exploration there is yet to be made into the remarkable human body. Why, for instance, should a meridian point on the sole of your foot be connected to a meridian in your bladder? A meridian point in your abdomen connected to a meridian point in a knee? What we do know is that gentle massage by our fingertips on meridian points gives us relief. You can get an idea of what meridian points are connected to what others by observing that when you get a twinge somewhere in your body, you may simultaneously feel a companionate twinge elsewhere in your body. I have a mildly arthritic area between my left thumb and index finger. When I massage its pain I sometimes feel electrical bursts of energy in my left big toe.

Could it be our meridians that make toe-picking so pleasurable for us humans? Whenever I sit around a swimming pool I have fun observing how many people are playing with, and massaging, their toes, usually unaware that they are doing so. Next time you are at a place where people's feet are bare, notice how many of them pick at their toes while they read, relax, talk, sunbathe. Perhaps you are doing it right this minute, while reading this book?

Beautification. Putting on makeup, grooming our hair, perfuming ourselves, all provide us with opportunities to feel the touch of a variety of substances, textures, and temperatures. The coolness of liquid face cleanser or makeup, the feel of a sponge wet with pancake makeup, the spray of a perfume atomizer pleasuring your olfactory sense receptors, the creaminess of lipstick, the pat on your cheeks as you apply rouge, the plastic or metal rollers in your hair, the touch of bobby pins, the feeling of metal, plastic, liquid, sponge, bristles of brushes brings a variety of sensations that our nervous systems enjoy.

When you're feeling down, moping around, and someone says, "Go put on makeup, you'll feel better," that's good advice. It is not only the color of the makeup on your face that helps to revitalize you, it is also the touch stimulation that your body gets while you are patting yourself with powder and rouge, brushing mascara onto eyelashes, drawing eyeliner onto upper and lower lids, smoothing on eye shadow. Hospital personnel know that patients feel better when they put on makeup. Human beings have been adorning their faces for a very long time. Esthetics were only part of their motivation for doing so. There was also innate wisdom that made them know that the touching of their faces was therapeutic.

We have other opportunities for self-touch sensations when we don a variety of jewelry on our ear lobes, necks, wrists, ankles, waists, bosoms, necklines. These activities call for small motor skills that give us pleasurable sensations on our fingertips. I enjoy the feel of screwing on earrings; there is something satisfying about making the tiny circles until the earring is snugly fastened. I like the feeling of completion when I snap shut the safety clasp on a brooch that I am pinning at my neckline.

I believe the increased use of cosmetics and the wearing of jewelry by men in recent years stem partly from the male's need to experience self-touch sensations. Traditionally men have been denied this kind of sensory input, except for those who are in show business and thus have occasion to wear makeup. The motions required to shave—the full palm and fingers moving as one unit in motions from the wrist—are different from the fine motions of the fingertips that women get to use much more than men do. We ought to encourage men to create more occasions, such as costume balls, for wearing makeup, so they too can experience the pleasurable self-touches of makeup and jewelry.

Sexual pleasure. Throughout history no human activity has been so widely practiced or so widely condemned, the cause of so much guilt, terror and fear, and yet the source of so much pleasure, as the simple act of stimulating one's own genitals to achieve release from sexual tension.

Masturbation is as natural a function as eating, sleeping and eliminating. It is the commonest way in which we humans express our sexuality during our early years. Anthropologists who have studied sexual behavior in many societies report that from infancy onward, as soon as children have muscular ability to do so, "most boys and girls progress from fingering of their genitals in the early years to systematic masturbation by the age of 6 to 8." Sex researcher Alfred Kinsey, in *Medical Aspects of Human Sexuality*, reported "orgasm-like responses in babies of both sexes at 4 to 5 months of age."[5]

To think of the horrendous things our forebears did to people simply because they enjoyed the pleasurable sensation of touching their genitals makes me angry. In the history of the human race, millions of men, women and helpless children have been beaten, tortured, imprisoned, tied up, because they put hand to genital. When I think of this I feel renewed zeal in my quest for acceptance of our sensual natures.

How did the powerful masturbation taboo begin? In early societies, rulers and religious leaders, wanting to increase their numbers and power, and to breed soldiers for armies of conquest, decreed under penalty of death that any sexual expression that did not lead to procreation was illegal and perverted. (This same proscription against nonprocreative sex also led to the outlawing of homosexuality and oral-genital sex.) There is a common misimpression that the biblical reference to Onan "spilling his seed upon the ground" was about masturbation; actually it was about *coitus interruptus* or withdrawal. Onan preferred to spill his seed rather than impregnate his brother's widow.

Through the ages, myths and superstitions have kept the masturbation taboo alive. Billions have suffered anguish because of warnings that they would go blind or insane, grow hair on their palms, never be able to bear a child, become impotent, suffer the loss of one's penis and pay a host of other terrifying penalties. Medical annals are full of cases of parents who, fanatically carrying on the taboo, burned children's hands by placing them on hot stoves, or tied together children's hands, to punish them for touching their genitals.[6]

These husband-and-wife reports to a counselor tell of their early experiences with touching themselves; they reflect what happened to many of us:

> THE WIFE: Mother would tell me I'd better be good. If I didn't quit, terrible things would happen. I'd be put away in an institution. Or she would have to be put away because I was driving her insane. I'd never grow up normal. I'd never be able to have children. I'd never lead a normal, happy life.
>
> THE HUSBAND: I remember at the age of perhaps four Mother told me that if I played with my teapot (a euphemism for penis), it would fall off. I have no memory of the act, only the constant warnings. As a clincher to her admonition, she cited the daughters of a neighboring family. They had lost their teapots for this very reason. Nonetheless, I continued to do what I was warned not to do. I'd sneak the forbidden act in and finish it as quickly as possible. Every time I did it, I was afraid the fateful day of losing my teapot would eventually dawn.

Now, happily for all of us, attitudes are changing. Dr. Karl Menninger of the Menninger Clinic says, "This metamorphosis in an almost universal social attitude [about masturbation] is more significant of the changed tempo, philosophy and morality of the twentieth century than any other phenomenon."[7] Masturbation had been something we did under threat of dire punishment. In the recent past sexual self-touching has begun to be approved as a reluctant recourse, something we were forced to do only if we didn't have a partner with whom to engage in intercourse. Now the growing attitude is that masturbation is something we do, *not instead of, but in addition to,* having sex with a partner. It is life-enhancing, one additional source of pleasure available to us.

When I recommend that people overcome their past conditioning and learn to masturbate, a good question arises: Can the electrical impulses traveling along our neural pathways be stimulated as well by our own fingers and hands as by another's? No scientific study has been done to determine whether the snap, pop, crackle, dash from synapse to synapse, described earlier, are as powerful when we touch ourselves as when we are touched by someone else. For me the answer is a resounding yes; I have experimented on my own body, stroking my breasts, arms, palms, shoulders, inner thighs. I have been astonished to learn with how much pleasure my senses have responded to my own touch.

Try a simple test on yourself, to learn the power you have to stimulate your own sensory receptors and to create feelings of joy and well-being in your own body. Lie down and relax on the floor of your home or office. You can be fully or partly clothed; if you are at home, you might prefer to be nude. If at this moment you are where you cannot appropriately lie down, you can do this test sitting up.

With your eyes closed, stroke your right index finger up your left thumb, down the other side of your thumb, and then up and down each of your fingers, slowly, slowly. Breathe deeply to ease bodily tensions and concentrate on what you are feeling as the receptors in your index finger happily meet with the receptors in your hand. When you have slowly done all five fingers, draw several small circles on your palm with your index finger.

By now you are probably feeling erotic sensations that are arousing and relaxing at the same time. Be aware of the intensity of your self-aroused sensations. Now move your finger up your arm, past your wrist, to the inside of your elbow and on up to your shoulder. Slowly, languidly. For many of you, this may be the first time you have ever explored your own body with such concentration. From your shoulder, stroke down the outside of your arm and back down to your hand.

Continuing to use the right index finger, slowly touch your hairline, your forehead, eyebrows, nose, cheeks, mouth. Stroke your lips several times. Which touch elicits the greatest response? Which the least?

Continue down your chin, neck and chest. Circle your breasts lightly, ever so lightly, and then your nipples. Your breath may be coming slower, deeper—like sexual breathing. Think of how marvelous your body is, with its 60,000 miles of blood vessels, five quarts of blood, muscles that enable you to perform incredible actions.

Continue tracing down your body to your abdomen, circle your navel, your pubic area. Enjoy the rising tide of sensation, welcome any sexual feelings, as you lightly circle your pubic area with your index finger. Then proceed down the inside of your left thigh to your knee and to the back of your knee, down the front of your calf and along the top of your foot, to the sole of your foot, back to your heel, and the Achilles tendon above the heel. As you draw your finger slowly up the back of your left calf, you may feel a responsive sensation in your inner thigh. Move your finger back up your body, and when your finger reaches your upper chest slowly let your right hand come to complete rest beside you.

Lie quietly savoring the sensations, the excitement and new knowledge you have just gotten about yourself. Freed from any inhibitions about touching yourself anywhere, you can picture the pleasure center in your brain smiling with delight. After a brief rest, use the index finger of your left hand to repeat the exercise on the right side of your body.

I have a theory that before humans can move up the evolutionary ladder to higher brain development, we must first take claim to our own bodies and become comfortable with every aspect of self-touch, including the sexual. That theory gains credibility from the reports of physiologists that the sense of touch is more acute in contemporary humankind than it was in primitive people—chiefly because we have larger and more recently developed parts of the brain in the neo-cortex. Studies of human origins suggest that it was when the neo-cortex began to develop that our fingers developed their sensitivities and amazing capacities. An enormous amount of our brain areas is committed to our fingers.

Charlotte Selver, teacher of sensory awareness, describes fingers as possessing "incredible magic." With these remarkable parts of your body, weighing perhaps a pound or so of your total weight, you have ten magic wands with which to pleasure, heal, beautify, nurture, relax and comfort yourself.

6

Touching Friends

During intermission at a dance recital at UCLA, a friend and I observed the physical contact among friends. Several hundred people stood talking in clusters of two, three, or more, on steps, walkways and the lawn. Astonishingly, in that throng, exactly two groups were touching. A little girl stood with her back snuggled up to a woman's front and the woman, presumably her mother, held the child close with both arms. On the steps a young woman playfully leaned against the back of her escort, with her arms around his shoulders, as he stood on the step below her.

All the others stood at right angles to one another, not looking at one another, making no physical contact, talking into the air. No wonder that we are described as an alienated society! Psychologist Sidney Jourard studied touch among Europeans by sitting in sidewalk cafés and other public places, clocking the number of times friends touched while talking. He reported an average of one hundred touches per hour. After a similar study in the United States, he reported that American friends touched two to three times an hour.

It is ironic. We go to great lengths to show friends we care about them. We send them flowery sentimental greeting cards or write loving letters. We invite them to dinner and devote exquisite care to the food we serve, the dessert we create, the beauty of the table setting, the quality of the liquor. We buy them presents that sometimes cost more than we can afford. We jubilantly share good news and console one another over bad news. We feel joy at the sight of them, especially when we haven't seen them for a while. *But we don't touch them.* We feel silly and self-conscious if we hold hands while we talk, or give a friend we love a hello stroke on the cheek. If we accidentally bump shoulders as two of us are walking together through a doorway, or if our thighs accidentally touch when we may be crowded into a car, we apologetically draw back inside our own bodies, sometimes even saying "Excuse me," as if we had done something rude. Ritualistic hello and goodbye pecks on the cheek, a cursory incomplete hug, or a handshake are all the physical contact we give one another. Columnist Ellen Goodman, in the *Los Angeles Times* points out that, in our "untouchable" code, buddies in movies "would die for each other, but never hug each other."

(I'm referring, of course, to nonsexual friendships. Most people will touch those with whom they share sex, or hope to—although many sex partners rarely touch outside bed.)

As we know, our resistance to touching stems in large part from our Puritanical antipleasure heritage. It may also derive, in smaller measure, from the fear of contamination that originated in the great epidemics of the past in which millions perished from bubonic plague, typhus and smallpox. "Don't touch, don't touch" was a common warning cry down through the ages, arising from contagious diseases. Our *homophobia*—"uneasiness about homosexuality"—certainly contributes to the gnawing anxiety that male friends, and some women friends, have about touching those of the same gender.

We are uneasy also because touch is a new language that we are just learning to speak; we have no dictionary to consult. We are uncertain both of how to "read" others touching of us and of how others may interpret our touching of them. We may be operating on the new wavelength of "touch is healthy communication," while others are operating on the old wavelength of "touch is sexy." We are only beginning to learn what the language of skin intimacy can say to us.

Consider this the beginnings of a dictionary of touch that you and I and all of society will be defining for a long time to come as we become more aware of the importance and effectiveness of touch in interpersonal communications. What touch says to us and how we respond to the communication depends on many factors—the gender of the one touching us, the body part being touched, the type of touch, the age of the toucher, the situation in which the touch occurs, our social status, and our level of intimacy.

When some people, while talking to you, touch your forearm, pat your upper arm, or give a light squeeze to your shoulder, do you feel good? Do you have a pleasant bodily response that makes you feel liked, admired, respected? Does the touch make you like yourself better? When others touch you, do you sometimes feel the touch as a demand upon you, a manipulation to get something out of you? Do you like physical contact from some people at some times and not at other times? Your varied responses reflect all these factors.

The same touch has different meanings, depending on the gender of the person with whom we are experiencing it. Let us say you are female and I am visiting in your city and we meet. I enjoy walking, so I might suggest we take a stroll. As we walk and talk, I might tuck my hand under your arm or draw your hand up under my arm. I love walking that way, especially because I have a tendency to walk fast and get ahead of the person I am with. People who walk arm in arm seem to synchronize their paces, which harmoniously affects them and the conversation. I also like the bulk of the other person's arm against my side; somehow it strengthens me.

You would likely accept my touch the way I meant it—an indication of friendly feelings toward you. You might even think, "What an outgoing warm person she is, easy to be with. I enjoy being with her."

(An American woman *might* be concerned about what passersby think. Might they misinterpret this as lesbian behavior? That would not concern many European women; it is social custom there for women—and men too—to walk arm in arm.) But if I were male and tucked your arm up under mine or put mine under yours, you might interpret my gesture as more seductive than friendly. A male who walks arm in arm with you is likely to project a romantic desire for intimacy, a hint of sexual potential. You could be wondering, Is he leading up to making a pass at me? By the slightest change in the pressure of your arm under mine, you would be letting me know through skin talk that you are not receptive; you might disengage your arm from mine as soon as you could politely do so. If you were receptive, you would probably increase the pressure of your arm against the side of my body, a signal of your interest and your encouragement of me to continue the contact.

Most of us follow widespread social norms that dictate which parts of our bodies are accessible to physical expressions of friendliness. At Purdue University researchers diagrammed the body, front and back, into eleven areas, to ascertain which parts could be acceptably touched by friends. As we would expect, the hands, forearms, upper arms, shoulders, head and forehead are more "accessible" for touch, both to males and females. But the genders differed on other body parts. Two studies reported in the *Journal of Communication* that women indicated that their thighs, lips, chests were not accessible for polite social touching, while men perceived touches to those parts of their bodies as friendly, warm and affectionate but not necessarily sexual. Women still believe—

and wish it were otherwise—that men touch mostly because they want sex, less to express affection or warmth or playfulness.[1] A powerful indication of how desperately men and women need to learn about touch is afforded by *The Hite Report on Male Sexuality*, which says that most men yearn for more non-sex-related touching from women![2]

A finding that surprised me was that both male and female did not consider a pat, tap or prod on their calves as suggesting sex. Many rated a calf pat as indicating "playfulness," "liking," or "friendliness." A woman, for example, might be walking around a swimming pool and a man might pat her calf in greeting. Both would understand it as a hello touch, not a sexual one. Another example: A man and woman are having dinner before going on to some event. It's time to leave. Either one could reach over and acceptably tap the other's calf, their friendly meaning being, "Come on, pal, let's shake a leg, otherwise we might be late."

If, instead of a brief patting or tapping of the woman's calf, the man around the pool gave her a long, slow feathery stroke down the length of her leg with his index finger, this gesture could communicate an entirely different meaning. It could be saying, "I'd like to be sexual with you." The *type or characteristics* of a touch—its pressure, repetition and duration—very much affect how we receive it.

To illustrate this point, let us go back to you and me—two women, taking a stroll when I am visiting your community. If you were to tell me of an achievement for which you were to receive a great honor and I patted your buttock, exclaiming, "Wow! How great!" I would be congratulating you by applauding you on your behind. My patting of that intimate part would be acceptable as long as it is laudatory. But if I were to repeat my patting of your behind a fraction too long, you would become uneasy, because its repetition and duration have now become ambiguous, evoking a different response on your nervous system—am I applauding you or being seductive?

The duration of a handclasp was cited in the trial of a lawmaker who was accused of having sex with underage girls. A young woman testified that she met the legislator when she was in the visitor's gallery, and he came over, took her hand and asked if he could be helpful in any way. Her immediate reaction was to accept the handclasp as routine politicking; but then, she said, there were some extra seconds of handholding that conveyed to her that he had fornication as well as friendliness in mind.

Our level of intimacy. Our friendships fall into three levels of intimacy: *acquaintanceship*, *friendly relations*, and *intimate friendships.*[3] People with whom you ride in an elevator in your building every day, the mailman, the cashier at the restaurant where you lunch, the bus driver who picks up your child, are all *acquaintances* with whom you have a low level of psychological intimacy. You may see one another often, perform acts of customary politeness, exchange greetings or comments on the weather, but you may not even know one another's names. On a rare occasion—the cashier happily announces that she is getting married or the mailman tells you he is retiring—you might clasp the hand of the acquaintance, or pat her on the arm, but you would feel it inappropriate to touch more intimately than that.

Our *friendly relationships* represent the largest block of people with whom we interact—co-workers, professional people (like doctors, dentists, counselors, teachers), those who belong to the religious, recreational or cultural groups that we belong to. We know one another's name and something about one another's life, family, feelings, beliefs, value systems. These relationships may often be pleasurable, fun, and sometimes interesting, but we feel no great obligation or responsibility toward them. These friends do not expect, nor do we want to give, much emotional involvement. If we don't see them for a while, we don't miss them. Certainly, if they were to ask for a favor that wasn't too difficult or too time-consuming, we would respond quickly. But if anyone did that too often, we might feel imposed upon and back off from the relationship.

With our friendly relationships we might touch arms, shake hands, embrace with an arm around a shoulder, or exchange hugs. Our touching of one another takes place mostly as part of hello and goodbye rituals, the touches being held a moment longer, our pressure a mite stronger, on those occasions when we haven't seen them for a while or when they may be leaving on a trip and we won't be seeing them for a while.

We *need* to keep most relationships in this friendly-relations category. It is physically and psychologically impossible to maintain intimate friendships with many people. We just don't have that much time available. We can give only a limited number of seconds and minutes per week to closeness, especially when we would also like to have our marital and blood relatives, spouses and children, among our intimate friendships.

Intimate friendships develop when two people move into deeper levels of psychological knowledge about each other. We tell each other our pains, hurts, disappointments, angers. Once we have such intimate knowledge of another person, obligation sets in. We must be available when an intimate friend needs to ventilate feelings of distress and to receive empathetic stroking from us. We make ourselves available to them on the telephone and in person. We feel needy of, and needed by, our intimate friends.

Because we need these people in our lives, we don't wait for fortuitous circumstances wherein we might both turn up at the same class or weekly bridge game. We make appointments to see each other, find time to talk on the telephone, and often travel distances just to be together.

When a person moves from the category of friendly relations into the category of intimate friendship, that almost always happens as a result of self-disclosure. Let us say you have a friendly relationship with a co-worker. You run into each other at lunch one day, and because the restaurant is crowded, you decide to share a table. As you talk, you develop warm feelings about the other person's empathy, good listening, humor. It feels good to be around her. You might then suggest that the two of you get together for a social activity on a weekend. During that activity you share disclosures about your spouses, lovers, children, life styles, career ambitions. Your *friendly relationship* is now moving into *intimate friendship.*

It is possible to move through the stages of acquaintanceship into intimate friendship in as short a span of time as one evening or one airplane ride! During the time it takes to fly across country, seat mates sometimes go from introducing themselves by name to disclosing matters about their marriages, their work, their parenthood, that they have never shared even with their spouses. I did so once on a cross-country flight. About to disembark, the woman and I hugged warmly. We knew that it was unlikely that we would ever meet again; and we might not want to, but we had had an intimate, emphathetic friendship during our five-hour flight, and it was appropriate to say goodbye with the intimate act of hugging, an acknowledgment that we had touched a deep core in each other. We often enjoy a one-time-only intimate friendship, because it provides us with a catharsis and yet does not carry the built-in obligation of such friendships when they are closer to home.

It is in our intimate friendships that most of us feel the lack of touch. Our psychological intimacy is not reflected in our physical intimacy. We may long to grab a close friend and tell her by touch language how we feel about her, but our antitouch conditioning holds us back. We feel an emptiness in ourselves, because we have not conveyed to beloved friends with the warmth of our beings how much they mean to us and how glad we are to have them in our lives. Some of us think, *Some day* I'm going to let her know how deeply I value her. I hope this book turns "some day" into "today" the next time you are with that friend. Then, you might walk arm in arm with her, hold her around the waist or by the hand, place an affectionate hand on her shoulder, give her a full frontal hug with a full-palm massage on her back (described in the chapter "Touching in Family Life"), or laugh with pleasure rather than apologize should you accidentally bump into each other getting into an elevator or walking through a doorway.

Your comfort in touching your friend will be determined partly by the *place* or *situation* in which the touch occurs. Let us go back to our imaginary evening together in which you and I might have shared confidences and become intimate friends. Winding up our pleasant time together, we might hug goodbye, standing in front of my hotel with other people around. But if we were in my hotel room we would make our hug brief—*intimacy feels safer in a crowded place or situation.* This was vividly brought home to me when I spoke at a meeting of the Los Angeles chapter of Women in Business, an organization of women executives and entrepreneurs. Forty women were seated in a circle. I was telling them of a fantasy that the child part of me has of being held close to the bosom of women friends in a mommy-child posture. The woman sitting next to me spontaneously grabbed me and held my head to her bosom as if saying lovingly, "Here, little girl, make your fantasy come true." We all laughed and I gave her a pleased embrace before disengaging myself to continue my talk. Her gesture was beautifully timed in that place and situation. Had the two of us been alone and I was telling her of my fantasy, I am sure she would not have grabbed my head intimately to her bosom.

How we feel about touching also depends on *social status.*

We have a strong caste system about who may acceptably touch whom. By society's value system, the person who is "dominant" because of wealth, achievement, job or social position often is the one who decides whether touch takes place.

I spend leisure time at a large apartment complex, where young Mexican men do the maintenance. I am friendly with a worker who has proudly shown me pictures of his wife and daughter. One Sunday afternoon a truck overturned, pinning him underneath. He suffered a broken leg and hip. As he lay moaning I comforted him by wiping sweat from his face and neck, stroking his arm, holding his hand, reassuring him help was on the way.

Months later he returned to work. When I saw him my first impulse was to hug him, welcoming him back. But I didn't; social status held me back. I felt that he might be embarrassed or made uncomfortable by an embrace from someone in a dominant social position. Instead I rubbed his shoulder and exclaimed how happy I was to see him fully healed. I felt angry both at the distorted societal values that prohibited my hugging him, and at myself for not flouting social conventions and doing what I felt like doing. That incident held a particular poignance, because I have a Mexican son-in-law. The next time I saw him I gave him an extra hard hug, wanting to make up for what I had not given his compatriot, wanting to apologize for all the ethnic prejudices in the world.

In a situation where I am the lesser person on the social scale, I touch friends and acquaintances in a manner that will help to elevate me to peer status. At psychology conferences and other professional meetings, I like to walk with my arm around the waist of colleagues, male or female. They feel the intimacy as I intend it—a nurturing gesture—and I enjoy feeling that even though I do not have the impressive diplomas and initials after my name that many of them have, I am enough of a peer to engage in this closeness.

Age affects how we respond to touching. An adult touching

a small child is acceptable if it is not a pounce or a rude distraction from whatever a child is absorbed in. Remember the "cuteness syndrome" mentioned earlier? We enjoy holding a small child, feeling its softness, its small-boned cuddliness, its unique sensual quality of "unused skin," as my daughter describes the feel of a baby. Its helplessness brings out the protector in us. When I hold a baby, I am inspired to fight harder to make the world a better place for that child to grow up in.

Adolescents are an age group that may be especially self-conscious about touching friends. Teen-agers yearn to touch and be touched. They fantasize and daydream about caressing and embracing. But with their sexual feelings constantly just below the surface, where touching can trigger an embarrassing erection or start vaginal secretions flowing, boys and girls often refrain from acting out their dreams. Instead, they engage in "surrogate" or "displacement" touches.

A boy would like to embrace a girl on whom he has a crush; instead he teases her with a mock punch to her upper arm or a pull on her hair. One day, at about age thirteen, my daughter came home from school in high excitement, shouting as she walked in, "Guess what, mom, a boy hit me in the playground today." "Mmmm, that means he likes you," I replied. She looked pleased. A girl's touch of a boy is disguised as "mommying" or "grooming." She'll reach over and neaten his shirt collar or straighten the hem of an outer shirt that is lying unevenly across his back.

Older people actually need advocates to fight for their right to physical closeness with friends. Again, because we confuse a desire for skin contact with a desire for sex we think that people past their procreative years ought not to have any such need or interest; there's something not quite "nice" about those who do. We still label elderly males "dirty old men," and we accuse older women who become friendly with young men of "robbing the cradle."

A California resort has an arrangement with a chain of

dance studios whereby it offers reduced rates for groups of patrons, who also pay the expenses of their dance teachers, so that, even if they are men and women without love relationships, they can still enjoy a weekend of dancing. Observing the joyful dancing of older women and their younger teachers doing the tango, cha-cha, rhumba, country western and disco, an onlooker muttered disapprovingly, "Why don't they act their age?" I hope my answer triggered a new attitude in her, a fresh way of looking at life. I said, "I admire these women for getting out and doing something about their needs, not sitting home feeling sorry for themselves or bemoaning their fate. I intend to do the same in later years."

A woman folk dancer has confided: "There's an elderly widower who holds me close whenever I happen to get him for a partner in a mixer (dances in which we change partners). I'm sure he's enjoying feeling my breasts against his body. There was a time I would have been outraged. Now I'm compassionate about his need to feel a woman's body close. So I let him hold me tighter than necessary for an instant. I hope that little bit of pleasure helps him to live longer."

I am aware that I like touching friends more than I like being touched. That's related, I'm sure, to a severe childhood burn that meant I was constantly handled as my painful injury was treated and bandaged. That trauma explains an irony in my life. I care very much about touching, but I don't like being massaged. Instead of relaxing me, the strong kneading, pushing, pulling, prodding, pounding of my flesh make me tense.

Nevertheless, I get great physical and psychological pleasure from touching friends, young friends, elderly friends, male friends, female friends. I feel as if I am polishing my skin when I touch people I care about. Recently I had dinner at a friend's home, along with her daughter and son-in-law. Afterward he stretched out, exhausted, on the sofa. I was sitting at one end of it. I put my hand on his ankle and instep and gave him a mini-massage, saying sympathetically, "You look drained." He responded with a sigh, pleased that someone was aware of his hard work.

Getting up from a booth in a coffee shop where I had breakfast with a beloved friend in her sixties, I kissed the back of her neck. It was so natural to do so. She's physically small, her head was bent over as she dug into a pocket, almost like a child, and I touched her with a kiss on the neck. Saying goodbye to a teen-age boy, a neighbor who was leaving to travel around the world, the two of us walked from my house out to his motorcycle, with me holding him around the shoulders. (When I ran into this young man several years later in New York, I was very moved to have him swoop on me with a big, affectionate bear hug.)

I am able to give these different kinds of touches with no embarrassment or awkwardness. I ask myself why? The answer is that my intent is nurturing and empathetic. It comes across the way I mean it, a caring touch that says "I hope this makes you feel better," or "It's a joy to be with you."

Even if my gesture were to be interpreted as having what Desmond Morris calls "minor sexual feelings," I don't care. I admit to having loving feelings toward much of humanity. My "minor sexual feelings" are transmitted into spiritual, not physical, arousal. I am turned on by spiritually erotic feelings for my elderly woman friend's bright mind and inspiring love for people; by the dedication of my girl friend's son-in-law as a social worker helping people; by the idealistic zest of my teen-age neighbor for learning about people in their native lands.

I have even felt "erotic" about the Jewish mamma of a waitress at Schwab's Drugstore who greets customers as if we were her children, doing everything but calling us "Tottela," rushing us morning coffee, worrying about whether we've had a good night's sleep, asking if we have something pleasurable to do that day. I feel like going behind the counter and hugging her, but I settle for giving her busy hand an affectionate squeeze. I feel turned on indeed about that elderly lady and the pleasure she gets from nurturing others.

Biologist Lewis Thomas, author of *Lives of a Cell: Notes of a Biology Watcher* and *The Medusa and the Snail*, urged a University of California graduating class not to hold back from each other the "affection and love" that many humans innately feel. We waste a lot of energy squelching and denying these feelings. A man named Bill Campbell wrote a letter to the *Los Angeles Times* about the sadness and nostalgia he felt at Beverly Hills High School's fiftieth birthday party. "It gave me people I wanted to hug," he wrote, "but was too shy to. One girl (or woman) almost reached forward for a fond hello kiss, and I almost responded automatically, but shyness prevailed. She will understand, I hope, that I wanted to embrace her and thank her for being one of those who gave me so many good memories, so good it hurts to think about them." I feel angry because Puritanical thou-shalt-nots deprive two people of the joy they might have gotten from holding each other close during a reunion after so many years! How wrong that their natural inclination to hug and kiss was so thwarted that they settled instead "for a somewhat awkward handshake."

This is one example of how we live with "massive inhibition of our nonsexual body intimacies," which applies "to relations with our parents and offspring (beware, Oedipus!), our siblings (beware, incest!), our close same-sex friends (beware, homosexuality!), our close opposite-sex friends (beware, adultery!), and our many casual friends (beware, promiscuity!)."[4] The word *promiscuous* comes from Latin; *pro* means "for," *miscuous* derives from *miscere*, "to mix." What lovely vibrations we could send winging through the world, what seismic energies we could transmit to each other if, in the real meaning of the word, we engaged in "pro-mix-cuous" affection by doing more touching.

I have had a taste of how energizing this can be. At a church workshop in sensory awareness, thirty-five men and women lay down in a row, spoon fashion. We lay quietly experiencing the hypnotic rise and fall of one another's breathing, feeling the cosmic force flowing through all of us. When the exercise was over, we jumped and danced like whirling dervishes.

For a remarkable experience, try this exercise at a party. You will be astonished, as I was, to experience the power of the energy we humans have available to give one another.

Inevitably, more of us are going to become more comfortable about touching. Technological facts of life will help to bring that about. As society becomes more computerized we shall be freed for shorter work weeks. Having more leisure, we shall be spending more time with large numbers of people, participating in sports, vacationing in camp sites, living in large residential complexes, attending classes, doing volunteer work in the community. We shall be drawn closer to some people than to others. We shall find ways to give special meaning to those relationships that we find more satisfying. One way in which we shall convey that meaning to those who are special to us is by gifting them with touch. Social forecaster John Naisbitt, in his book, *Megatrends*, calls this "high tech/high touch." The more technology we live with, the more we are going to need each other's touch.

A friend of mine lives in a Florida compex with 15,000 residents in adjoining apartments. She and her special friend, a widow many years her senior, find to their surprise that they are physically affectionate. They hug, kiss and hold hands. Such intimacies grew naturally out of their frequent contact, the proximity of their apartments, their nurturing of each other at times of stress and their happy grabs at each other at times of joy.

If, like most of us, you have been holding back because of embarrassment or shyness, when your natural instincts are to embrace and touch friends, how do you free yourself so that you can enjoy the benefits and taste the delicious "flavors" of touch? Here is a step-by-step primer designed to condition you gradually to the agreeable feelings that Nature meant you to have, instead of the anxious feelings that society conditioned into you. You will literally be making new recordings on your nervous system to help you feel good about touch.

1. Discuss this chapter with friends whom you want to touch.
2. When you are sitting and talking, touch a friend often with a slight pressure on the forearm, upper arm or shoulder.
3. When you are walking, put your arm through a friend's arm or draw her hand under your arm. Hold your arm-in-arm posture a short while, but not so long that it becomes tiring.
4. When a friend appears tired, say so in a nurturing tone of voice and rub his or her upper shoulders, neck and back. Just do it; don't ask, "Would you like me to rub your back?" People sometimes want it and say no out of embarrassment.
5. On an appropriate occasion, put your arm around a friend's waist and walk a while that way. Don't keep your arm there so long that it begins to weigh heavily on your friend's back.
6. On a stroll take a friend's hand and hold it for a little while.
7. Sitting at a table after a meal, dawdling over coffee or dessert, take your friend's hand and hold it a short while.
8. Touch a friend's cheek in hello or goodbye with a loving open palm as if you were a nurturing mother greeting a child. The child part of every one of us likes some babying.
9. Stroke the back of a friend's head as you're saying

"Thank you" after you've enjoyed hospitality at her home.

10. Remember this Nigerian proverb. Hold a true friend with both your hands.

When you initiate touch, your friends will respond gladly. They have the same secret and inhibited yearnings for closeness that you do. Here is how a writer, Paul Chance, reacted to a friend's touch. He was finishing an interview with Evelyn Hooker, a distinguished researcher into homosexuality. Reporting on ten hours of conversation about her work, Chance concluded: "As we parted, she hugged me; what's more startling, I hugged her back. You have to understand that I am rather a stuffy, formal fellow, the kind who would shake hands with his own children. But Hooker, warm and gentle without reservation, can rip the seams in any stuffed shirt, even mine."[5]

7

Touching Strangers

Sociologist Erving Goffman, who has spent years studying people's behavior in public places, believes that everything a stranger does in public is made up of tacit threat or promise. A woman may glance at us with a happy look on her face or walk by with a joyful stride, and her behavior holds out promise for us too. We get a lift when we witness a stranger's joy. A stranger gesticulates with irritation while talking to someone and threatens our psychic harmony. An angry person, even a stranger whose anger is not directed at us, makes us less comfortable within ourselves.

Recently, in the supermarket, I observed two men in a check-out line. One courteously moved his basket out of the way of shoppers trying to get through the crowded aisle. The other, apparently thinking he wanted to usurp his place, instantly became furious. The two exchanged angry words. Feeling diminished by the scene, I went home distressed that some people are so quick to do battle. This behavior of strangers was a threat to the peace of those who witnessed it.

We feel threatened by strangers far more frequently than

we feel promise. That is because threats, even though implied and not actual, restimulate early childhood learning. "Stay away from strangers" and "Don't talk to strangers" are deeply recorded admonitions. We go through life almost *expecting* dreadful experiences with strangers. Long after the need for childhood caution is past, this *xenophobia*—"fear of strangers"—is reinforced by the violent times in which we live. As grownups we still go around asking ourselves the ancient feudal question: "Who goes there, friend or foe?" (Ironically, more than half the attacks and murders in this country are not of stranger against stranger, but of family member against family member.)

Arthur Schopenhauer, the German philosopher, once compared humans to hedgehogs in winter. Our problem, he said, was to position ourselves close enough for warmth but far enough apart to escape being pricked by each other's quills. Every person of goodwill must feel this conflict. We live with an ideal that we want to help create a better society. We want to trust and be friendly with most people, including strangers. But the harsh facts of life hold us back.

I take a stroll and I would like to give out messages of friendliness by my demeanor and facial expression, but I'm afraid to. Is that lone male walking toward me a potential mugger or rapist? Is that teen-age female walking rapidly behind me a possible purse snatcher? I don't like myself for feeling this way. What a relief it could be to live in a time when we need not be so suspicious of one another. It is exhausting to switch back and forth from foe to friend mentality. How much more energy we would be able to keep for ourselves if we didn't have to expend so much of it being phobic about strangers.

If we were conditioned not to be fearful of strangers but to act on our instincts, how might we behave?

In experiments at Swarthmore College, men and women, all strangers, stayed from one hour to an hour and a half in a room that was pitch black except for a tiny pinpoint of red light over the door in case a subject got panicky and had to leave. No one did. The research subjects could do anything or nothing—whatever they wished. They did not know the purpose of the study. Their actions were photographed with infrared cameras.

Freed of social constraints, what did seventy strangers do? After a few tentative moments in darkness, a threatening setting in which you would expect people to huddle into themselves and avoid contact with others, almost 90 percent of them began to touch each other on purpose, seeking comfort and security in the feel and warmth of one another's body. Some touched hands and arms; some hugged; some stroked faces. Some sat in groups with their bodies touching. The researchers reported that "with the simple subtraction of light, a group of perfect strangers moved within approximately thirty minutes to a stage of intimacy often not attained in years of normal acquaintanceship." Many subjects thoroughly enjoyed the experience of touching and being touched by strangers and said they would participate in such an experience again. It felt so good to reach into dark, barren space and encounter warm flesh and responsive strokes and squeezes. The researchers concluded that "if the social norms governing our relationships did not keep distance among us, the sharing of intimacy such as in the dark room would be widespread."[1]

Not only is it *desirable* that we develop a positive credo about strangers, it is *imperative* that we do so, for we are destined to spend much of our lives interacting with them. According to human-relations specialist Eli Ginzberg of Columbia University, "Pretty soon we are all going to be a metropolitan-type people in this country without ties or commitments to long-time friends and neighbors."

On the day you read this, tens of thousands of us are in transit from one home or apartment, one community or state, to another. Tens of thousands are leaving old jobs and starting new ones. Every year, millions of us change the places where we live and work.

In addition to the moves that Americans are making, millions more men, women and children are moving here from other lands to join us. Detroit has the largest Arab population of any American city. Los Angeles has the largest Mexican, Bolivian, Ecuadorian populations outside those people's native lands. Other cities have growing numbers of Vietnamese, Thais, Koreans, Cubans, Iranians.

Before long, if it is not already happening, we will scarcely go a day without encountering new co-workers or schoolmates, new neighbors in our apartment house or on our block, new faces in the company elevator or cafeteria. Every one of us needs to define a personal ethic of trust enabling us to feel comfortable with all these strangers. Touch can be a valuable social tool to help us dispel our fear of one another.

Every one of us knows the difference between spending our first uncertain day in a new situation with people who greet us with a polite but distant smile and with people who touch our hand, arm or shoulder and say, "Welcome to the company" or "Welcome to the school." I attended many schools. The uneasy young stranger I was would have welcomed any touches of greeting as I went through my first anxious days in fourteen different schools.

We need a credo of trust, so that the diversity of people whose paths cross ours can be a source of joy and not a threat, and so that they can know that we, who are also strangers, are not a threat to them. My credo comes from John Howard Lawson, an historian and screenwriter who, as one of the Hollywood Ten, was jailed during the McCarthy era for his defense of the Bill of Rights. He told a church audience, in a Sunday morning sermon, that he would "rather live trusting nearly everybody and risk being taken advantage of by someone once in a while than live by distrusting everybody."

There is no more effective way for strangers to convey trust and potential friendliness than by the reassuring act of touching. It is better than words; words *say* the message, but touching *acts out* the message. The warmth, the satisfying pressure, the feel of another, becomes magical. A stranger does not know us, yet trusts us enough, thinks well enough of us, to make physical contact with us. What a boost for our self-esteem. We like ourselves and one another more.

Purdue University researchers confirm what most of us know intuitively—touching makes strangers happier and more positive about themselves and others. Librarians were asked alternately to touch and not touch the hands of students as they handed back their library cards. The experimenters then interviewed the students. Those who had been touched reported far greater positive feelings about themselves, the library and the clerks than did those who had not been touched. The difference in their good feelings was significant, even though the touch lasted only about a half second and half the students didn't even remember having been touched. The fulfillment we get from skin contact is amazing; even when we don't know we are being touched, it still has the capacity to make us feel good.

"Who's got the time to go around touching strangers?" you may ask. "I hardly have time to touch friends and loved ones." We Americans certainly are obsessed with time. As we have seen, we ask the same question about delivering babies and about touching our families. Touching strangers doesn't require extra time. Almost invariably it happens while we are talking to them, greeting them, helping them. A carful of tourists stopped on Sunset Boulevard to ask me directions. I gave the driver the information, indicating with one hand the direction he should go. With my other hand, I gave his forearm a slight "good luck, have fun" squeeze. That touch was both seen and sensed by the passengers, who all smiled and waved thanks as the car drove off. It didn't take an iota of extra time.

But it sent a lot of people off in good spirits, including me. I felt like a good human being.

These people were what I call *staying* strangers, those with whom we stay in physical proximity anywhere from a few seconds to several hours. For a small block of time we are together in a joint place and situation. This is one of three categories of stranger we encounter throughout life.

Think over a day in your life. In what activities do you interact with staying strangers? A day in my life might involve me with such strangers in line at the post office, browsing at the library, trying on clothes in the large public dressing room of a discount clothing store, occupying adjoining seats at a lecture. We don't know the names of staying strangers or anything about them. We would feel silly if we went around exchanging names, nor is there any reason to do so. Imagine the reaction of a woman half undressed in a try-on room if I were suddenly to say, "My name's Helen Colton. What's yours?" Probably we shall never see such strangers again and we wouldn't recognize them if we did.

With some staying strangers we inevitably share unintentional body contact. Your body may be crushed against theirs in a jammed elevator. You may be on a crowded subway or bus, your hands unavoidably touching as you hang on to the same strap. Intimate parts of you, such as buttocks or breasts, may be touching strangers as you push past them to get to the exit. We would be embarrassed or uncomfortable to have similar intimacy with friends. We can tolerate it with strangers, because there is no obligation to converse with them or to have any ongoing interaction. Anonymity sometimes brings liberation.

An advertising poster for a record company shows bodies jam-packed on a bus, with the large bosom of a female straphanger almost resting atop the head of a young woman, seated in front of her, whose face is all but pressed against the straphanger's abdomen by the pressure of the crowd. Other passengers are pressed against the straphanger's buttocks and back. While the poster is exaggerated for humor's sake, it nevertheless sums up the kind of situation in which population density frequently forces staying strangers into intimate bodily contact.

Another category of stranger—*joining* stranger—includes those with whom we engage in a mutual activity once, occasionally, or regularly. They could be people in our gym class, tennis club, church choir, backgammon tournament. Or someone with whom we chat at a social gathering. Whatever the setting, we have some connectedness with joining strangers during a block of time, even if we may not exchange names. At folk-dance festivals I dance with hundreds of people. We hold hands, we twirl in each other's arms, but many of us never know the names of those we have danced with.

Throughout our lives we constantly go through joining-strangers situations. For single men and women, this frequently repeated occurrence, at singles discussion parties, dances and mixers, is often a cause of psychic pain and unease. "How I hate entering that room alone that first minute or two," single clients tell me. (Despite the facts that we have growing numbers who choose to remain permanently single and that singles will one day outnumber marrieds, society still places a unique "value" on twosomes. The implication at present is that you are a less valuable human being if you are alone.) Because of my own discomfort when I was newly single, I worked out a technique that helped me get over my loneliness and to make touch contact in an acceptable way with other joining strangers. Here's what I learned to do.

On entering a room, take a seat not far from the door. As a woman enters alone, smile at her, pat the seat next to you, and say, "Would you like to join me?" The woman, who is probably also dreading being alone, would respond gratefully. Introduce yourself, ask her name, and make physical contact by shaking hands. As a few others enter, looking as if they are newcomers and don't know anyone, invite them to join you. Ask each newcomer his or her name, greet that person with a handshake or a light welcoming touch on the arm, and introduce him or her to the others in your small group. Clients confirm that this extemporaneous reception does for them exactly what it did for me—it earns warm vibrations of gratitude for helping a few joining strangers through the awkward part of a singles evening before they go on to circulating comfortably.

The third category is that of *passing* stranger, people who pass us on the sidewalk, in airports, shopping malls, at the beach. We don't touch, because there is no reason to do so, and it could be misinterpreted as antagonistic or aggressive. What we do give to one another is "civil inattention," an odd form of politeness in which, as we pass, we avert our glances and do not look into each other's eyes, a way of not invading each other's privacy.

Great sexist inequality exists in "civil inattention." A male can look openly into the face of a passing female and feel safe. A female does not feel equally safe looking openly into the face of a male, who is likely to see it as a sexual come-on. Women are forced to give a lot more civil inattention to men, averting not only their eyes but often their faces. As a jogger I especially resent this inequality. I often avert my gaze from a strange man's face as he is looking directly at mine. I do not have as much freedom of the sidewalk as a male does.

In some situations we may quickly go from one category of strangerhood to another. Let us say that a teen-ager is riding a bicycle past you on the sidewalk. Suddenly he takes a bad flop, and you stop to offer help. You and he would then become staying strangers for a little while. You might stay with him long enough to help him up, to make sure that he is not seriously hurt, or if he is, you might go to telephone for help and then stay with him until aid arrived. During your brief experience as staying strangers, you could appropriately touch him with a comforting pat, by holding him under the armpit to help him up, perhaps by wiping blood from a bruise. Studies of strangers helping strangers show that we are more likely to help someone in distress if we are the *only* person witnessing the incident.[2] When others are around, we become ambivalent about whether others will help, and we are less apt to take action.

All our friendships begin when joining strangers get together. Think back to the occasions when you first met your closest friends. Did you meet as joining strangers when someone introduced you at a social occasion? Or while engaging in a mutual sport or hobby; or in a classroom, at work, at a boat marina, in the building or neighborhood where you live? The more time we spend with a joining stranger, the safer we feel in exchanging friendly touches. Our touches become physical evidence of our growing trust in each other.

Even in the most tactilely liberated society, it is unlikely that people will go around touching passing strangers. An anthropologist friend conjectures that someday, perhaps centuries from now, that too may change. He says hopefully:

> In an abundant world where everyone's needs are well met with ample space, food, comforts and security for all, there will be no impetus for strangers to be fearful of one another. It is an economy of scarcity that creates the fears we now have. It could be the custom for passing strangers who feel like it to give each other little "love pats" of touch, like strokes of recognition.

Until that Utopian day, we need to use discretion in touching strangers. A major factor in how a stranger reacts to our touch is our physical position at the moment of contact. How would you feel if a stranger suddenly came up behind you without warning, clapped you on the back and shouted, "You've just won the state lottery"? For an instant, before responding to the good news, you would feel angry, because the stranger's sudden touch triggered the uncomfortable startle reflex (as we've seen, Nature's way of preparing you for "fight or flight"). You might think, That jerk! for scaring you, and then be annoyed at yourself because "that jerk" is bringing you wonderful news.

A sensible rule is never to touch a stranger (or a friend, for that matter) from the rear unless he has first seen you in front of him. Touching from behind carries an implied threat of attack. At a party once I sat rubbing the back of my neck. A stranger came up behind me and suddenly put his hands around my neck and shoulders. It felt as if I were about to be choked. I screamed with fright. He apologized—"I saw you rubbing your neck and I was going to give you a massage." If he had not scared me by approaching in back, I would have enjoyed the warmth of his hands. Instead I was enraged at him for doing such a stupid thing.

Industrialist Justin Dart once saw former president Gerald Ford ahead of him on an airplane. He came up and clapped Ford on the back. Ford, whose nervous system must be especially sensitized to sudden motions as a result of the assassination attempts on his life, went white and jumped. Turning, he saw Dart, whom he knew. We can speculate about what went on in Ford's mind for that brief instant before he saw who it was.

Our reactions are different, however, in a situation in which there is a reasonable *expectation* of being pushed or jostled from the rear by strangers—say on a crowded commuter train, where you expect people to push past you to reach the exit. Your expectations alert you to that kind of bodily contact and you are not startled by it.

In what situations might we accept and even enjoy being touched by strangers?

Sharing happy news. The days on which armistices are signed are those on which much touching of strangers occurs. Overwhelmed by the good news, we throw normal cautions to the wind and hug, kiss, jump up and down with strangers. At sports events strangers hug and thump each other in excitement when a score is made, as mentioned in the chapter "Touching in Sports." Afterward some feel sheepish because their high spirits have overcome their wariness of strangers.

Suggesting a pleasurable experience to come. Coming out of a movie with a friend who was "high" on the comedy we had just seen, she spontaneously touched the arms of strangers waiting in line to get in and gleefully announced: "You're in for a treat. It's a funny movie." They responded with grins of anticipation and amusement at her childlike enthusiasm.

Imparting useful or nurturing information. After Halloween, a florist gave away pumpkins that he had used for display. I was delighted when a stranger in a drugstore across the street touched me on the arm, happily pointed to the pumpkin he was carrying, and said, "You can get one for free over there at the florist's." Most of us are grateful for a stranger's thoughtfulness in telling us where to save money or get something for free.

Sharing an experience in a carnival or play setting. At the Laguna Art Festival I sat resting on a wall, my back touching the back of a woman facing the opposite way. Neither of us made a move to disengage our backs. The festive atmosphere in which we expect strangers to be in a friendly, holiday mood made our back closeness mutually tolerable. Elsewhere we would probably have said "excuse me" and moved away.

Our uniqueness as a stranger. When Lillian Carter, Jimmy Carter's mother, was a Peace Corps worker in India, she walked through a village where "all the people would come up to me and touch me—it unnerved me at first, but I learned to grin and bear it." As an elderly American woman, she was a rare sight to the villagers.[3] Anthropologists on field studies often wait for tribespeople to come to them and make touch contact. They are glad when that happens. The touches tell them they are trusted and may now proceed to further interaction.

Lunda Gill, an ethnographic artist who paints portraits of remote tribes, says in *People* magazine, "I'm an oddity and they are leery of me at first. Many times they bring out their children to touch me because they have never been exposed to a white person and have never seen blond hair before."

Comforting at times of tragedy. In such a situation all prior attitudes are suspended. The best in human nature comes to the fore. That is because an innate part of the human personality likes to nurture others. Also, when strangers are experiencing tragedy we suspend our fear of them. At such a time, a stranger may go spontaneously to a distraught parent who is quivering with shock because her child has been hit by a car, embrace her and give her bodily comfort and solace.

I have been observing a great change in the touch behavior of television interviewers and newscasters. In the past, these people would maintain journalistic neutrality, hiding their emotions when they were dealing with people suffering tragedy. Now TV talk-show host Phil Donahue often clasps the hand or touches the arm of people who are hurting. He held hands with a mother telling about finding the body of her son, a suicide. Henry Alfaro, a Los Angeles newscaster, was obviously moved as he touched a small boy whose face had been grotesquely burned and was being rebuilt through plastic surgery. Connie Chung, Los Angeles TV newswoman, reached out consolingly and squeezed the arm of a grief-stricken Salvadoran mother, who was telling of the brutal killing of her son.

These behaviors of admired media personalities are indications of changing social attitudes toward bodily contact between strangers. In a sense, they give permission to viewers to do the same—to be tender and tactile with hurting strangers.

Congratulating one another for having escaped disaster. Thomas Wolfe, in his novel *You Can't Go Home Again,* wrote about a fire in a large apartment house in which escaping tenants who had never before spoken or nodded hello now gathered in the courtyard and "were soon laughing and talking together with the familiarity of long acquaintance." A prostitute comforts an elderly tenant by wrapping her chinchilla coat around the woman's shoulders.

Admiring a child or pet. What easier way is there to start communicating with a stranger than to be around a child or a pet? We melt inside when we see an adorable toddler or puppy or kitten. We have an urge to feel its softness and warmth. While I was standing in line at the bank, an enchanting little boy with merry eyes began to play hide-and-seek with the grownups. A moment before, we had all been dour strangers. Suddenly several of us were chatting, touching each other in conversation as we reminisced about our own children doing that, asking the child's mother questions about him, kneeling to talk to him and pat him.

Touching in absentia. When you walk into a restaurant or a waiting room and you are cold, notice how it feels to sit in a chair just vacated by someone you don't know. I love when that happens, it feels delicious! I think, Ah-h-h, thank you, whoever you are, for giving me your body heat.

In some situations it is wise to be wary of touching strangers. Our touch can cause us to be sued, land us in jail, harm us, or even kill us! Delicious though it feels to embrace a cuddly child, it is foolish to do so when it might be misinterpreted as child molestation. I saw an irate father make a citizen's arrest of a man fondling his two-year-old daughter. The man protested that he had no malicious intent, he was just so pleased by the child's beauty, but the father insisted on calling a policeman, who drove the chastened toucher to jail.

Our decent instincts make us want to help an injured stranger. However, Good Samaritan Laws state that you can be sued for moving the victim of an accident. Unknowingly, you can cause a fractured bone to pierce a vital body part. It may be better just to provide a warm cover and summon professional aid quickly. It is unfortunate that such well-intentioned laws add to alienation and to bystanders' alleged indifference to human suffering. I once heard a man say at the scene of an accident, "I'd like to help, but I'm afraid I'll do the wrong thing and be sued." Better laws need to be written protecting the bystander and also the victim he would like to help.

Mentally or emotionally disturbed people, especially psychotics, sometimes cannot bear to be touched; they rear back in fright or lash out in rage. They would benefit from being cuddled and stroked, but it needs to be done in a safe environment such as a therapy session. Or touching may be okay if *they* do it and are in charge. But a stranger's touch may elicit violent behavior. This is probably the underlying cause of incidents we read about. In Ohio two fathers were in line with their children to see Santa Claus. One father accidentally shoved the other father, who pulled out a knife and stabbed the man to death before horrified onlookers. On a Los Angeles corner two motorists banged fenders. As they stood angrily exchanging names and license numbers, one jabbed the other with a finger to his chest. The man who was jabbed drew a gun and shot the other to death. He pleaded in court—"It was only when he poked me, I lost control and pulled out my gun."

Because we have great numbers of disturbed and potentially violent people walking our streets, we need to redefine what we mean by *bravery* and *cowardice.* In my value system it is bravery, not cowardice, to walk away from such a person. He may not have the controls on his behavior that most of us do, and so we may not be dealing with a normal conscience. Confronting such a person is not brave, often it is foolish, a no-win situation. When I said this during a guest sermon at the Burbank Unitarian-Universalist Fellowship, a couple came to me afterward and thanked me. The wife had tears in her eyes as her husband said, "I've been wanting to confront a neighbor who seems to be a disturbed person. Now, after what you've said, I'm giving up on that idea and I don't have to feel like a coward anymore."

How can lay people, untrained to recognize disturbed behavior, protect themselves? Unfortunately, there is no general rule; we have to go partly on intuition. It is sensible not to initiate strong or prolonged contact with any one who we suspect or know is disturbed—he could misinterpret it as an attempt to harm him—but to be friendly and accepting if that person offers touch to us. Mental-health workers are usually trained to say, "I'd like to touch you (or hug you or hold you). Is that okay with you?"

A Mexican friend, Esther deGally, after attending a conference of rural women from Latin America, wrote a long narrative poem about her moving experience with strangers. One verse, inspired especially by Caribbean women, says:

> Embracing often and warmly
> their body expression overwhelmed our wary
> protectiveness.
> Of what?
> Freedom grew a new face.

My own freedom is growing a new face. Writing this chapter has been clarifying and therapeutic for me. It has helped my new ethic evolve. I have lessened the number of strangers of whom I am wary. I've been helped by a scholarly treatise, *The Stranger: A Study in Social Relationships*, by Margaret Mary Wood, Ph.D., in which she emphasizes the distinction between "types" of people and "individuals."[4] Many of us, as I did, fear whole *types* when we may actually be fearful of *individuals* who fit into those types.

I have had inordinate fear of one type—young males wearing knit caps and leather jackets. This was irrational bigotry. I have never had an experience with any such individual. I know that skiers, ice-skaters, janitors, delivery men and other males wear such clothing. I have undoubtedly been influenced by television and film stereotyping of such garments as being worn by potentially violent youth gangs. The vast majority of young males wearing knit caps and leather jackets are not muggers or attackers. Nor are the majority of any type of stranger. Learning not to categorize *individuals* as *types* diminishes my fears.

We pay a high price for our fears. Because we are conditioned into keeping our distance from strangers, our loving relationships suffer. To do "social battle in the streets and offices" we don "heavy suits of emotional armor" inside of which we distance ourselves physically and emotionally. We find it increasingly difficult to shed this armor and become intimate and tactile with those we care about.

I have a notion that where an individual might be thinking of perpetrating an antisocial act, such as a mugging, an action or touch from a potential victim showing trust in that individual might possibly work to deflect an attack. Perhaps by conveying to a potential assailant the suggestion that the person is trustworthy—is in fact trusted—we can diffuse hostile intentions. I certainly am not suggesting that we go around touching possible attackers, just that we consider a possibility that, in some situations, trusting behavior might result in an individual's split-second decision to *be* trustworthy and not commit a contemplated act.

A colleague had an experience that may validate my notion. She was walking at twilight after a business appointment in a high-crime area. She heard several young men behind her. Fear crawled up her spine. Making an instant decision she turned, faced three youths and said, "Can you gentlemen please tell me how to get to Clark Place?" All three began to give her directions. Touching the nearest one lightly on the arm she said, "Thank you all so much. You've been most helpful," and she walked rapidly to the other side of the street. We certainly don't know whether the men had intended to harm her. That could have happened spontaneously, without forethought, as they got closer to a tempting target. By implying her trust in strangers, she evoked helpful—not harmful—behavior from them.

I regret that these perilous times require us to take precautions. I still avoid dark streets, and I sometimes carry a change purse in a pocket rather than in a handbag on my arm, but I no longer fearfully skulk through a walk. Reducing the number and types of strangers of whom I am afraid, I live with a lighter heart. I like to imagine that, in small ways, I am already living in a better, more trusting society of the future.

When that better society does come, what behavioral signals will strangers give one another? These will be among the social clues:

People will smile more at each other, reflecting their sense of well-being and of trust.

They will perform more frequent acts of courtesy and concern. Years later I remember the man who, when I was struggling across a parking lot pushing a baby in a stroller with one hand and pulling a basket of groceries with the other, said, "Here, I'll help you," and pushed the groceries to my car.

They will touch us and be touched by us.

Social psychologist Stanley Milgram, discussing our willingness to be close to strangers as a result of some disaster, asks, "Is there any way to promote solidarity (among strangers) without having to rely on emergencies and crises?"[5]

Those of us with pioneering spirit have the power right now to start bringing that ideal to reality. We like people whose touch makes us feel good. If each of us were to touch a stranger once a day in an acceptable manner and place we would be holding out tacit promise rather than tacit threat. We would be responsible for one less unit of fear and hostility in the world. Each time we touch a stranger, we give that person implied permission to pass a touch on to another. Our touch can work toward creating a more trusting, less fearful, society.

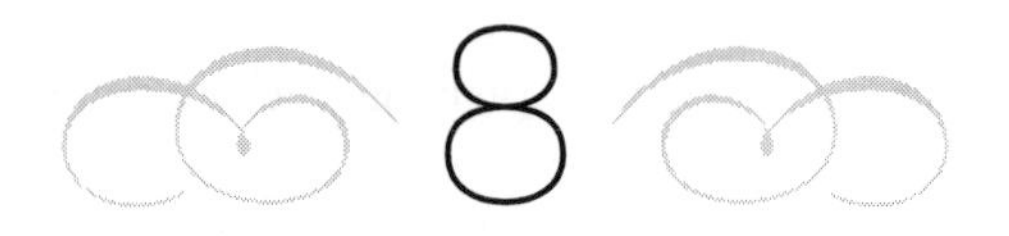

8

Touching in Illness and Grief

The last thing I remembered as I was being wheeled to the operating room was the comforting holding of my hands by my two doctors. They had been walking down the hallway on either side of the cart on which I was lying, each of them stroking a forearm and hand as I drifted deeper into unconsciousness. I could no longer see or hear them. The reassuring pressure, the protective bulk of their hands on mine, was my last sensation as I went off into the black void of surgery.

How I needed their touches! It was August 1951. Ninety days earlier, biopsies on tumors on both breasts had showed up benign. Now I was going back into surgery. Many more tumors had developed. Both doctors were deeply concerned that this time the lab tests might show cancer. I could awaken with both breasts gone.

It was a hellish time. I fought going back into the hospital and almost had to be physically dragged there. The cancer taboo was more powerful at that time; *cancer* was a dirty word. To have cancer was to feel almost criminal, to bear a secret shame. I was distraught when I entered surgery, and I shall always be grateful for the only comfort I knew—the surgeons' reassuring stroking that somehow gave me hope. These doctors were not indifferent men of medicine efficiently performing their art. They were human beings who felt deeply about me, who ached for my misery and fear, who ignored the widespread dogma that doctor should not touch patient except when absolutely necessary during medical procedures. When I awakened later from the anesthetic they whooped with joy, "Good news, Helen," and patted my face endearingly, announcing that my tumors, once again, were benign. (I've never had a recurrence.)

These two medical men were in the minority then, and they are still in the minority a third of a century later. During illness, a time when we more desperately than ever need the nutrient of touch, the medical profession too often still writes "N.P.C."—No Physical Contact—on patients' charts. These are coded instructions to the staffs in hospitals and convalescent homes. A nurse at a Veterans Administration hospital reports this horror story resulting from an N.P.C. notation on the chart of a helpless Vietnam veteran. Due to the high turnover of personnel, all of whom perfunctorily looked at the chart and continued the instructions written by the first physician, the young man lay for weeks untouched except for feedings and occasional quick sponge baths. It finally dawned on a new doctor that the veteran was wasting away. He asked the nurse to give the veteran a gentle massage. As she began to stroke his body, "his skin shed like a snake's," she reported with horror. "It actually sloughed off in my hands from lack of stimulation."

Doctors may not always write the N.P.C. message but they often act it out during their hurried, machinelike daily rounds. To observe many a doctor making his rounds is to be appalled at a "healer's" frequent indifference to a patient's hunger for a reassuring squeeze on the arm, a pat on the back, or a handclasp. Norman Cousins, in *Anatomy of an Ilness*, describes the "utter void created by the longing—ineradicable, unremitting, pervasive—for warmth of human contact. A warm smile and an outstretched hand were valued even above the offerings of modern science, but the latter were far more accessible than the former."[1]

Some doctors treat disease; they do not treat people. Anthropologist Edward T. Hall once heard a nurse describe physicians as "beside-the-bed doctors, who were interested in the patient, and foot-of-the-bed doctors, who were interested in the patient's condition."[2]

Certainly in a few situations it might be harmful to touch a patient, such as a burn victim whose healing skin could become infected. But in many cases absolutely no reason exists for such an instruction as N.P.C., other than the mindless, ritualistic perpetuation of a tradition that, in view of what we now know about the therapeutic value of touch, is indefensibly detrimental.

At Grant Hospital in Columbus, Ohio, a nurse, Pamela McCoy, validated the theory that touching speeds recuperation. Twenty patients were touched, twenty others were not touched. She reported in *RN Magazine* that among those touched there was a "vast reduction" in the incidence of "complaining, angry, disgusted, uncooperative patients." Eighty-five percent of the touched patients made positive responses about the hospital and its personnel, and recuperated faster than the untouched patients.[3] Touched patients typically make such statements as "They're lovely to you here."

How did the no-touch tradition among doctors begin? It probably stems in part from the Hippocratic Oath which Hippocrates, the ancient Greek father of medicine, imposed upon his disciples as a code of ethics. Among the paragraphs of this oath is the phrase "I will abstain . . . from seduction of female, or male, bond or free." It is customary for medical-school graduating classes to recite this oath. As we have repeatedly seen, touching has been equated with sexiness and seductiveness. To avoid any hint of seduction, physicians avoid touching except for the "necessary medical procedures," especially when a patient is undressed for an examination. This proscription against touching has been reinforced through the centuries by concern about lack of sanitation and fear of spreading disease. However, in these days of antibiotics, sterilization of equipment, sanitation and other means of controlling disease, physicians no longer need fear doing harm by touching. The potential good from a physician's friendly touch may more than compensate for any possible harm.

The no-touch attitude has been reinforced by Freudian psychoanalysis, the major influence in mental health for decades after Freud defined his theory of repressed infantile sexuality as the source of much of our mental and emotional disturbance. Freudian followers have felt, and some still do, that touching a patient only deepens the problems arising from sexuality. It is significant that willingness to touch among doctors is growing at the same time as adherence to Freudian principles is waning. Some psychoanalysts insist that Freud never intended to proscribe tactile contact between doctor and patient. Psychoanalyst Maurice Walsh says, "This is a misunderstanding of Freud by some poorly trained analysts. Rigidity about touch has no place in psychoanalysis."

Whatever the original occasion for the touch taboo and the subsequent reasons for perpetuating it, it is no longer necessary or desirable that physicians observe it. Some doctors protest that they avoid touching, because they are afraid of being thought seductive. But no patient is going to yell "Rape" if, when her doctor is saying goodbye after an examination and consultation, he puts his hand on her clothed upper back, upper arm or shoulder and gives a slight pressure meant to comfort and reassure. My mother felt ten feet tall when her admired Dr. Joseph Echikson would put an arm around her shoulder as he was saying goodbye and ask: "Rose, how is Sonny (my father's nickname) behaving?" That this busy and important man should take time to make this contact made her feel valued.

Much doctor-patient interaction occurs in situations where doctors would not be vulnerable to cries of "rape" or "seduction." Furthermore, as a precaution against accusations, nurses are often present in examining rooms when patients are undressed. Seductive doctors and patients don't even need to be alone with each other, and touch is not necessary, to hint at sexual intent. That can happen through tone of voice, by prolonged eye-to-eye contact, through double-entendre words the two may exchange.

While the numbers of seductive doctor-patient relationships are not great, the publicity about them is. As a result, the fear of touching permeates the medical profession even in situations that could not possibly have sexual implications. A television series, *Lifeline*, showed real doctors in real situations with real people. Patients, families and doctors often went through hell together, but they never touched. On one program, a doctor attended a woman through the long, agonizing birth of twins. Afterward he said to the exhausted mother, "They're going to take you to the recovery room now." He put his two hands out toward her as if he were going to comfort her with a touch, a sign that he was aware of what she had just gone through. Instead he placed his hands on the railing of her bed.

On another program a woman pediatrician went through more than thirty days of anguish with a couple who were waiting to see whether their premature baby would live or die. He died. As the doctor walked the grieving couple to their car to say goodbye, I literally ached with the hope that she would make physical contact with the mother. "Hug! Hug!" I kept yelling inwardly. "As two women who've shared such sadness, you must hug." The mother got into the car. The doctor shut the door and waved the couple off. I felt ill.

Critics of the medical profession maintain that it continues the tradition of not touching because it considers itself an elite class, superior to the rest of us. Medical schools have been accused of promoting elitist attitudes that encourage doctors not to make physical contact with the mass of lower beings—us—except when absolutely necessary. Says one critic embittered at standoffish doctors, "To touch is to humanize. Most of the medical profession like to keep their dehumanizing distance." Another critic, coining a word, says they have "pedestalized" their profession high above the rest of us. "Actually they are not a profession, they're a business," he says, citing the growing numbers of doctors who incorporate themselves and who own pharmacies, laboratories and medical buildings.

A doctor friend tells me that it is less the medical-school indoctrination that keeps a physician from touching than it is "his or her own personal style. Medical schools don't teach you to touch," he says, "but they also don't tell you *not* to touch. Some doctors are speculum and stethoscope warmers and some are not." He was referring to an "inside the profession" criterion for judging "softies." Some doctors are indifferent to the unpleasant feel of a cold instrument on warm skin; they apply a cold stethoscope to a warm chest or insert a cold speculum into a warm vagina. You probably have had such an uncomfortable experience and it made you flinch for an instant as you reacted to the thermal shock. (The feminist movement includes "warming of speculums" among the changes it wants doctors to make in their treatment of women.) Doctors who care about patients' comfort take time to warm cold instruments under hot running water before applying them to the body.

Next time you visit a doctor observe whether he or she touches you other than during the examination. There is a marked difference between the doctor merely saying, "You can get dressed now, then come into my office," and the doctor patting your shoulder while saying it. Sir William Osler, the noted Canadian physician, a pioneer fighter for the therapeutic value of touch, said, "Taking a lady's hand gives her confidence in her physician."

The Chinese people have a saying, "Fate decreed that people must die of starvation. Fate decreed that girl babies were worthless. Fate decreed there must be wars. *Fate must go!*" I believe *tradition must go!* Any tradition that no longer contributes to humanity's physical, mental, psychological or spiritual enhancement has served its time and should be discarded or updated—especially the one about the medical profession not touching patients. There are hopeful signs. Third-year students at the prestigious University of California Medical Center at San Francisco now take a course "on the human side of the practice of medicine, to help them understand that being sensitive and responsive is an essential feature" of their careers. Students are learning "how the appropriate approach, attitude and *bedside manner* [the emphasis is mine] can accelerate healing." A Los Angeles physician whose practice is devoted to cancer patients says, "I always touch them, always, always. That is very important." Significantly, the morning of the day I wrote this, a newscaster reported that the American Medical Association has chided medical schools for failing to teach future doctors more humanistic treatment of elderly patients.

At Brown University School of Medicine, medical students participate in training sessions exploring their attitudes toward touch. Actors portray patients in real cases; students play themselves as future doctors. In a personal letter, Dr. Stephen R. Smith, Assistant Dean of Medicine, reports this incident:

> The case was that of an 81-year-old widow with a traumatic ulcer of the leg. The student-doctor was explaining to the

> actress-patient her need for near absolute bed rest. This raised the subject of dependency which was upsetting to the patient, who prided herself on her fierce independence.
>
> Reviewing the videotape of the encounter, it became apparent at a point of great emotional intensity that the student wished to reach out in a gesture of emphathy and consolation, but he withdrew.
>
> The videotape was interrupted, and the student was asked what his feelings were at that moment. He said he had wanted to touch the patient but wasn't sure if that was appropriate. We asked the actress to tell what the "patient" was feeling. She replied that she felt very much in need of warmth and human support. Nothing would have been better than having the doctor just hold her hand for a few moments. The student's hesitation and reluctance was due to his own uncertainty about what was appropriate in the practice of medicine. Never before in his training had he been told that it was okay to touch patients for non-diagnostic purposes. By hearing from the "patient" that this would have been a welcome touch and from his instructors that this was not only okay but desirable, a new horizon was opened for this student.

The branch of medicine that may be the most resistant to change is mental health. Undoubtedly this is so because it is also the branch that has had the highest incidence of doctor-patient sex relations. A book widely used as a text for psychotherapists is still advising them that "it goes without saying that physical contact with the patient is absolutely taboo."[4]

A therapist whom I went to years ago kept this book on his desk as his bible. Not only did he not make physical contact, but he forbade those in his group to do so. Among the reasons I care about touching is my soul-searing experience in a group session that he led. He had the poor judgment to keep me on the "hot seat" for two hours, at a time when my ego strength was incredibly low. I shudder at the memory of how desperately I needed someone to give me body comfort with an embrace, reassurance that I was still okay and a good person despite the verbal abuse. I got none. If I had, it would not have taken the several months it did for me to recover my equilibrium about my worthiness. That experience led to my becoming a mental-health counselor; I knew that there had to be a better way than the one I experienced, so barren of comfort to suffering people who are seeking insight.

One day we will look back upon that textbook sentence with the same horror as that with which we now view bloodletting. Now we are beginning to know, in the words of Rudolf Arnheim, art philosopher at Harvard University, that "Our senses are a source of sanity."[5]

There is no law written for all eternity that says the Hippocratic Oath, enunciated 2,500 years ago, may not be updated to fit contemporary humanity. Hippocrates also said, "Some patients recover their health simply through their contentment with the goodness of the physician." In *Modern Medicine Magazine*, Bernard Lown, professor of cardiology at Harvard University School of Public Health, deplores that "physicians are too busy with a laying-on of tools. Both presence and touch help to establish a reassuring connection with the patient." I hope the initials N.P.C. on patients' charts and in physicians' behavior will be replaced by T.L.C.—taking those initials one step beyond "Tender, Loving Care," to include "Touching, Loving Care."

Just what happens in the bodies of sick people to make them feel better when they are touched? To learn this, Dolores Krieger, R.N., Ph.D., professor of nursing at New York University, has studied the effects of the laying-on of hands. Her hypothesis, that treatment by the laying-on of hands would significantly alter hemoglobin levels in the blood of an ill person, has been repeatedly confirmed. She also found that the toucher had to have a "fairly healthy body" and had to care very much about helping the patient. Based upon her studies of Eastern religions and ancient healing techniques, she realized that *prana*, a Sanskrit word translating into "vitality," or "vigor," in English, was the energy subsystem underlying healing. The healer literally "invigorates" the patient by stimulating the level of hemoglobin, the pigment of the red blood cells which carries oxygen to the tissues. As tissues receive increased oxygen, they are energized to carry on the regenerative process. Krieger says, "I know that when I lay my hands on or near an ill person, he has a subjective sense of heat in the area that is ill or diseased, a sense of relaxation and well-being, and his hemoglobin values change following treatment."[6] (A healer, Krieger notes, may also achieve results not necessarily by making body contact but by moving her hands slightly above the painful areas.)

Touch may even be *the* basic need of patients, more vital than medication. When a patient's need for touch is satisfied, he or she is strengthened and better able to deal with problems and traumas. It may be that lack of touch has contributed to a patient's illness in the first place. At Vanderbilt Hospital, Nashville, seventy-five psychiatric patients were asked whether they preferred to hold or be held by others. They favored being held over holding. Why? In their daily lives many of them had been expected to be nurturers of others. Among the causes of their breakdowns was that they needed to get what they were always expected to give—physical comforting, holding, touching.[7]

Ashley Montagu reports an astonishing cure from massage.[8] A thirty-year-old woman had suffered severe asthma attacks all her life. Montagu's suspicion that she had been deprived of touch in infancy was confirmed when she told him that her mother had died at her birth. Suspecting a relationship between development of the respiratory system and early tactile stimulation, Montagu suggested that she attend a physiotherapy clinic where she would be massaged to compensate for her early deprivation. During several months of massage treatment and since then, July 1948, the woman has never suffered "a single serious asthmatic episode." Montagu's research further shows that a shoulder hug, "putting one's arm around the shoulders" of an asthmatic person, may sometimes alleviate or end an attack.

Some mental-health practitioners believe that schizophrenia may have some roots in touch deprivation in infancy and childhood. One method of treating schizophrenics is to "reparent" them—diapering, feeding, holding, bathing, caressing them as if they were infants and gradually moving them up through stages of psychological development into adulthood, constantly providing the bodily contact that they lacked in their original parenting.

When a psychotic person is going through an episode in which he is out of control, ranting, hallucinating, thrashing out at others, therapists may cuddle him as if he were a baby, stroking and rocking him back into relaxation and quietude. Anyone who has observed a psychotic patient subside from frenzied screams and sobs into whimpers and moans and then into sighs of calm relaxation, entirely through being held, caressed and rocked, can never doubt the power of touch to heal. I have observed this awesome experience several times, and I am always deeply moved by the capacity of our wonderful hands to bring surcease to a troubled soul. (Compare this touch treatment with "olden days," when such patients, so needy of intimate body contact, were immobilized in straitjackets and placed in solitary confinement!)

Touch has also been found helpful in retarding the deterioration of senility. Researchers at the University of Georgia studied forty-two people who were seventy or more years old and were living in a nursing home. They found a correlation between "sensory deficits" and senile traits such as irritability, forgetfulness, and careless grooming or eating habits. The elderly who received massages, frequent stroking, hugs, squeezes of the hands and arms, love pats on the cheeks and affectionate touches of their heads, showed fewer signs of senility. They were more alert, good-humored and physically vital.[9]

Dr. James J. Lynch of the University of Maryland and Dr. Aaron Ketcher of the University of Pennsylvania, in the October, 1982 *Sixty Minutes* report, "Man's Best Medicine," stated that their research showed cardiac patients who had pets to hug and caress outlived those who had no pets. The stroking of pets actually lowered patients' blood pressure.

There are growing numbers of humanistic people, especially among younger doctors and psychologists, who are working to get the medical establishment to accept touching as part of healing. In the forefront of this emerging group are David Bresler, Ph.D., former director of UCLA's Pain Control Clinic and founder of the Center for Integral Medicine (12401 Wilshire Boulevard, West Los Angeles, Calif. 90025) and Dennis Jaffe, Ph.D., author of *Healing from Within.*[10] It was Bresler who gave a widely quoted prescription for a woman patient to bring home to her husband: four hugs a day. To the husband's surprise, the prescription worked; the patient had needed caresses, not capsules.

The future will see growing acceptance of the medical specialty of touch practitioner, one trained in touching, hugging, massaging, acupressure and other types of tactile therapy, who work on prescription by physicians, just as occupational and physiotherapists do now. (If you wonder why people have to be trained in how to hug, go back to my chapter "Touching in Family Life.") Often, in noncritical illness, touching may precede drugs as the treatment of first choice. Physician William Steiger says, "We should offer warmth and succor *before* we order the first tests." Many patients will undoubtedly get well without pills or potions. (One can foresee the powerful pharmaceutical lobbyists politicking against touch as therapy, which would cut down on the profitable pill-popping of our overmedicated society.) So, some day, if your doctor, after examining you, sends you into an adjoining room to get hugged by a touch practitioner, don't resist—go on in and enjoy it.

By nurse Dolores Krieger's standards there must be millions of natural healers among us—loving, compassionate people who have healthy bodies and whose electrical energies and physical warmth can help others. Think of the positive effect on men and women who now feel unneeded, who could awaken each morning with a sense of mission, going to hospitals, sanitariums and convalescent homes to share their body warmth, stroking and holding the sick.

Mary Ohno, nursing specialist at Queen's Medical Center in Honolulu, says sitter-touchers are urgently needed for those who have had cataract or other eye surgery and have both eyes bandaged. Living in darkness, "eye-patched patients" cannot watch a hospital's daily routine. They hear speech coming from disembodied voices whose faces they cannot see. They easily become disoriented, hallucinate, and show other signs of sensory deprivation. Lay people could help by staying nearby and frequently touching such patients.

Throughout the land we have hundreds of thousands of bars with Happy Hours to give people alcoholic release from tensions. Why don't we have an adult version of Laura Huxley's Project Caress for infants, mentioned in Chapter 1? Why not Happy Hours in the recreation rooms of apartment complexes, gymnasiums, health clinics and therapy centers to which needy people can come for the touch prescription of being embraced and rocked?

Many physicians and nurses are aware of the power of touch, but they are reluctant to use or prescribe methods that can be attacked as unscientific. In *Human Behavior Magazine*, James J. Lynch writes that even people in deep comas often show improved heart rates while their hands are being held. He urges the helping professions to overcome their reticence to accept "medicine beyond science."

Ordinarily we do not think of grief as an "ailment." However, as far as our bodily responses are concerned, grieving *is* a kind of sickness, or it may be the cause of our becoming ill. Dr. Thomas H. Holmes and Dr. Richard H. Rahe devised a stress scale, which assigns units of stress to forty-three good and bad events in our lives. (Not all stress is negative; Hans Selye, the pioneer in defining human stress, says that we have positive stress that energizes us for play, fun, and joyous experiences.) The units range, for example, from 100 for the death of a spouse through 29 units for change in work responsibilities, as in promotion, demotion or transfer, down to 11 units for minor law violations such as a traffic ticket. If we accumulate 150 bad stress units within two years, there is a 33 percent chance of our coming down with a major illness or undergoing an accident.

We may experience a major loss—death of a loved one, loss of a job or a love relationship, not receiving an anticipated inheritance, having a home destroyed by fire, flood or tornado—or undergo a less disastrous grief, such as failing an examination or losing a cherished watch inherited from a parent. Whether it is a major or a minor event that causes our grief, our bodies respond in the same manner. Our glands work overtime pouring out hormones, we may undergo adrenal exhaustion, our immune system is weakened, and we may become ill.

When our minds are hurting, it is highly therapeutic for our bodies to have a warm body and shoulder against which to press our faces and sob out or talk out our pain. Another person's body heat is soothing, counteracting the cold and chill, the hollowness, we feel when we grieve. Being enfolded in another's arms shields us against more psychic assaults at the moment. We feel childlike, protected. Someone is sharing our burden of sadness; that lightens the load. Being held can mean the difference between going into a depression, turning inward our anger and hurt about our loss, or achieving a catharsis through discharging our pent-up anguish, enabling us to cope better.

When Robert Kennedy was shot in a Los Angeles hotel, author Theodore White went at once to the bedroom of Kennedy's thirteen-year-old son to give him "bodily comfort." James Roosevelt, eldest son of Eleanor and Franklin, tells how his mother, after the death of her brother, "spent her hurt in Father's embrace."[11] A poignant incident occurred in a classroom at the University of Southern California, when a student afflicted with osteosarcoma arrived after a grueling session with his doctors, who had just told him that his body was not responding to treatment and his illness would be fatal. Grief-stricken, he unashamedly asked his classmates to hold him. They did. What those students learned about human needs that sad day, as they took turns holding their friend, was more powerful than anything they might have learned in a textbook.

Even when we are not the one who is held, but the one doing the holding, physical contact helps us to cope better. A newborn was on a life-support system at UCLA Medical Center. There was no hope for the infant. Authorities and parents agreed to remove the system. Before it was disconnected, the weeping mother and father cradled their baby between them for twenty minutes, finding solace in the baby's warmth before its body should turn cold. Phil Donahue asked the mother of Karen Quinlan, who had been in a coma for years, a question that a segment of the population was voicing, "Why don't you let her die?" Ms. Quinlan said, "Because I can still see her and touch her."

Dr. Lynch, mentioned earlier, says:

> I have been struck by the way that most people finally say goodbye (to loved ones who are terminally ill). They will speak to each other, if the patient is physically able, usually in subdued tones; they will try to make every effort to ap-

> pear confident, and sometimes they will even joke. But when they say goodbye, it is almost as if some deep, primitive, instinctive ritual takes over. . . . Just before leaving, they will stop speaking and silently hold the patient's hand or touch his body or even stand at the foot of the bed and hold the patient's foot.[12]

Words are inadequate at that moment; the final goodbye is one of touch.

Touching the sick and the grieving, we must learn what degree of pressure and how prolonged a pressure is comfortable for them. When a person has a fever or serious illness, too much pressure hurts. I learned a lesson that gives me a twinge of remorse. I gave a hearty handshake to a man who had leukemia. I instantly realized my blunder when he turned white and began to sweat from the pain of my strong pressure. I now know a light, airy touch is needed. A widowed friend told me that at her husband's funeral, a friend standing behind her during the long graveside rites kept her hand on the widow's shoulder. "Her hand began to feel like a weight of iron. I kept wishing she'd take it away. She meant to be consoling, but she caused me physical pain."

Touching at times of sickness and grief is a measure of a culture's humanity. If we do not give one another those elemental nourishments—our body heat, the light pressure of our hands, the "velvet quality of satisfaction" from skin contact—when we are most vulnerable, can we consider ourselves civilized? Animals lick and comfort their injured. Kung Bushmen of Africa, in a custom called the Trance Cure, dance themselves into a healing trance, working up a great sweat so they can lay their warm, moist palms on the sick. We withhold touch and tell our sick and injured not to cry. A play, *For Colored Girls Who Have Considered Suicide When The Rainbow Is Enuf*, ends on a heartrending note, with a woman crying out her deep, tragically unheard yearning: "I want the laying-on of hands, the laying-on of hands." That spine-chilling wail of hurting people echoes down through the centuries. Is it a primitive cry calling us back to our basic natures? Long before the human race had pills, we had palms.

Touching a Person with a Disability

I originally entitled this chapter "Touching A Handicapped Person." Then, while attending a meeting of the American Booksellers' Association at the Los Angeles Convention Center, I got into a conversation with a woman in a wheelchair and asked if I could interview her. "I'm not handicapped," she said. Seeing her shriveled arms and legs, I looked at her questioningly. "I am a person who has a physical disability," she said. "All of me, the total person, is not handicapped. I think, I talk, I read, I feel, I laugh, I cry, I love, I like, I dislike, I chew, I swallow, I eliminate, I'm curious, I like movies and drives into the country, I sing—off key; but I sing."

I got her point. I reached out and squeezed her arm, acknowledging that I was grateful to her for giving me a new perspective. Our touch was warm skin to warm skin. Nothing in that connecting of our bodies indicated that some of her muscles were withered and useless; skin cells and receptors in both of us were alive and vital. We broke into laughter of pleasure that two people, strangers a moment ago, had so quickly achieved rapport.

Along with most of us, I had been unconsciously equating a disability in a part of the body with disability of the total person—if muscles in your legs, arms, back or neck are not functioning, then *you* are not functioning. And, like most of the nondisabled I had been equating *physical* disability with *emotional* disability—if your torso is numb, without feelings, so is your psyche; therefore you cannot be suffering hurt or pain if we ignore you, avoid your eyes, and do not make touch contact with you.

America, like India, has its "untouchables," but in this country it is based not on social class but on physique. The disabled consider themselves an outcast group living behind "barriers of architecture and of attitude," people whom we, the rest of the world, would just as soon keep hidden because their presence in a world designed for the "perfect" physical specimen makes us feel ashamed and guilty that we have our wholeness and they don't. Describing prejudice toward the disabled as another "ism" along with sexism, ageism and racism, a disabled man calls it "physicalism." We all have a vested interest in doing away with these barriers. As millions of us live into our eighties, nineties, hundreds and beyond, many will have temporary or permanent infirmities that come with aging or we will be dealing with infirmities in our families. We saw poignant evidence of this when Senator Barry Goldwater, once an athlete and active sportsman, hobbled on crutches to the podium at the 1980 Republican Party convention, after surgery for calcium deposits in his hip. By doing away with barriers for *them* (the disabled), we will be doing away with barriers for *us* (the disabled of the future). Disabled people jestingly refer to us as "*T.A.B.s*"—temporarily able-bodied.

Who are the disabled? Why are we so uncomfortable making social and physical contact with them? They have been estimated as including anywhere from twenty to fifty million persons. Uncertainty arises from the fact that until 1970, the United States Census never even included a question about disability. We don't have an impression of that many disabled among us; that's because many have "invisible" conditions such as deafness, impaired vision or physical immobility caused by rheumatism and arthritis, which the rest of us don't see. Many with "visible" disabilities, such as the cerebral-palsied, the brain-damaged and the mentally retarded, are in institutions unseen by the able-bodied population. Others with "visible" disabilities—the crippled, the amputees and the spinal-cord-injured—are in wheelchairs and do not have access to public buildings and transportation, so they spend much time trapped indoors or in their home neighborhoods. A *New York Times* reporter, Sonny Kleinfield, walked from Lower Manhattan to Central Park and back, and did not see even one person in a wheelchair![1]

Nearly all disabilities are caused by one or more of the following:

- Birth defects and trauma such as cerebral palsy, mental retardation, blindness, deafness, congenital conditions. Some years ago it was the medical custom, no longer followed, to put premature babies into incubators with a high oxygen content; many were blinded for life. Tens of thousands of adults throughout the world have been born deformed, some with missing arms and legs, some with hands attached to their shoulders like flippers, because their mothers were given thalidomide during pregnancy.
- Illness, such as measles, rubella and scarlet fever, and diseases of the nervous system, polio, multiple sclerosis, muscular dystrophy. It was an attack of high fever from an undiagnosed illness (biographers conjecture that it may have been scarlet fever or meningitis) that left Helen Keller deaf, mute and blind when she was less than two years old.
- Military injuries during war and during peace.

• Home accidents such as falls in bathtubs, off roofs and ladders, on slippery floors and down steps.

• Aging, which may bring strokes, rheumatism, arthritis, arteriosclerosis, osteoporosis.

• Accidents at work, such as that experienced by a woman training to become an airline stewardess who broke her back sliding down an exit chute and is permanently paralyzed. Nearly a half million farm workers suffer disabilities every year. Tractors turning over and pinning a driver underneath are the greatest source of permanent, farm-related injuries, according to L. Dale Baker, safety engineer at Cornell University's College of Agriculture and Life Sciences.

• Accidents with guns. Behind that two-inch newspaper item about a child shot by a playmate or a sibling there may be a lifetime of paralysis for the victim. Such tragedies are happening more often as more private citizens keep guns.

• Accidents in cars, planes, trains, on motorcycles, bicycles and skateboards. These cause the largest number of severe injuries and paralysis.

• Injuries in sports. I had known about injuries on the football field and on ski trails, such as the fall that paralyzed ski champion Jill Kinmont, but I have been shocked to learn how many young people are crippled because of water sports. James Shepherd, whose parents founded the Shepherd Spinal Center in Atlanta following their son's injury, became paralyzed while surfing off Rio de Janiero when a monstrous wave slammed him against the ocean floor and twisted his spine. A San Diego teen-ager, horsing around on the beach during a Boy Scout outing, dove into a pond on a dare. The sound of his head hitting wet sand and the crack of his spine breaking into pieces haunts him. He is in a wheelchair for life.

Disabled people have a great variety of immobilities. The cerebral-palsied may walk or limp but be unable to control muscles required for talking or eating. Their minds may be brilliant, but they may be unable to convey their thoughts in speech. What agonies of frustration must be suffered by some cerebral-palsied who have brilliant minds trapped in helpless bodies! Psychological literature is full of cases of children who have been "warehoused" in institutions as retarded, until tests reveal that they have high I.Q.s. A palsied man whose tortured speech is barely understandable struggles to speak these poetic words describing his rolling, lurching gait: "I feel as if I am skating over marbles."

Some disabled do not have muscle power for gross motor activities; they cannot raise an arm, bend an elbow, move their heads. But they can work muscles controlling fine motor activities—their fingers move and they can write or type, turn knobs and dials, turn a page, hold eating utensils and feed themselves with motions from their wrists. Some retain gross motor skills and can propel wheelchairs with their arms but have no finger dexterity—they cannot grip a fork or spoon, hold a comb, or manage zippers or buttons, so they must be fed and dressed by others.

Even though they are not able to move muscles from the neck down, some disabled persons perform amazing feats of typing, painting, sculpting and writing by maneuvering wands held in their mouths or attached to bands around their foreheads. Sara Needham, a Woodstock, New York, artist who was born with no legs, no hands, and only one arm up to the elbow, sculpts huge stone works with tools held in her mouth. She is lifted to the tall structures and placed on a scaffold by her husband or by friends.

Just by blowing their breath, some can control pneumatic devices to start gas ovens, roll wheelchairs, turn lights on and off, activate television sets, radios, alarm clocks, air-conditioners and electric doors. Some use typewriters and tape recorders by pressing keys with their tongues.

When that vitally important "apposable" thumb (see Chapter 1) is paralyzed, a person cannot grasp. To get an idea of how this feels, hold your thumbs inside your palms and unscrew a jar top using only your four fingers on each hand. Accomplishing this task becomes maddeningly awkward and difficult. In Nature's design, every muscle and joint has its purpose; when even one joint is immobilized, we are impeded in our tasks of reaching, grasping, holding, opening, closing, lifting, carrying, setting down, walking. When my daughter was a baby I was dancing with her in my arms when my right little toe hit the edge of a sofa and was broken. I hobbled for three weeks until it healed, protesting all the while—"How can one little toe make so much difference?"

Each day, sadly, the number of disabled increases. Every twenty-four hours, thirty men, women, boys and girls—10,000 a year—become paralyzed for life as a result of auto accidents. If their immobility is such that they cannot move their arms, they will never again be able to reach out and enclasp a loved one.

Our discomfort with the crippled and the blind is born out of the ignorance of antiquity. Ancient people believed a deformed person was "a child of the devil," a punishment from God. A poor soul would suffer not only the physical pain of her affliction but also the psychological pain of being stigmatized by everyone because she was frighteningly "different." I vividly recall that, as a child, I was horrified when I watched films with biblical backgrounds in which village people would point fingers of scorn and shriek "Leper, leper" at creatures whose agonized expressions tore my heart out.

We still carry on the ancient custom of stigmatizing those who are different. We may express neutral-to-positive attitudes toward the disabled because it is socially "nice" and "proper" to do so, but when we are in their presence, our eye-movement patterns, sweating and heart rate, reveal discomfort, anxiety, avoidance of eye and tactile contact, and other physical signs of our negative feelings. Robert Kleck of Dartmouth University asked college students to enter a room and start a conversation with the person inside it. Sometimes he described the person as an epileptic; sometimes he didn't. The students sat farther away when they thought the person was epileptic. Kleck performed the same experiment using a fake amputee. He got the same results. Students distanced themselves from the "cripple."[2]

A subliminal reason for not wanting to be around the disabled is our collective guilt. They are a constant reminder that with all our technology we cannot "heal" them. We can repair the "nervous system" of a rocket or a missile, but not the nervous system of a person with multiple sclerosis or muscular dystrophy. Satellites can scan the earth and read the license number on a car five miles below but we cannot bring vision to the blind child of a mother who had measles during her pregnancy. We have created the technology to travel millions of miles into space but not to travel the fraction of an inch that can fuse two broken pieces of spine back together.

Because we feel ashamed of ourselves, we often get children to act out our negative feelings for us. Children can be openly cruel toward those who are "different" or "not normal."[3] On the Phil Donahue Show, singer Bette Midler told of a tragic experience. Her closest chum in high school in Hawaii was a crippled boy, who committed suicide because he couldn't endure the taunts of other students. A crippled girl tells of her life:

> When . . . I began to walk out alone in the streets of our town, I found that wherever I had to pass three or four children together on the sidewalk, they would shout at me. Sometimes they even ran after me, shouting and jeering. This was something I didn't know how to face, and it seemed as if I couldn't bear it. For a while those encounters in the street filled me with a cold dread of all unknown children.

Both children and adults practice the silent cruelty of staring. A polio victim recalls this painful experience:

> I stand in the doorway to a restaurant, my heart sinking, my body filling with a great tumor of anguish. The sensation is akin to both swelling and shrinking. I wish that I would shrivel away to nothing, like a paper curling to ashes in the grate, or that I would burst with the swelling.
>
> People! Eyes! Eyes!
>
> Something in the back of my mind moans, Oh God, O God, O God! I lash my courage: Enter! Get it over with!
>
> But you need not see eyes; you can feel them. Even when you know that you are past them and they are behind you, they reach out. . . .
>
> The aisle is miles and miles in length but at last I reach my table and seat myself. I relax. There is a light dew of sweat on my face. I'll be all right in a minute. The tablecloth is long, my legs are under the table. Now to order, eat, and enjoy a hard-earned meal. I don't want to look too far ahead. After a while I must get up and walk out again.[4]

The first time I read these words I was sitting on the outdoor terrace at the Los Angeles Music Center, waiting to go in to a matinee. I raised a newspaper in front of my face to hide my tears from passersby. Now, weeks later, typing these words on a Sunday afternoon, I am once again overcome by emotion. I ask myself why I have such a strong reaction. My tears are for my mother, blind from diabetes during her last thirty years, my own "tumor of anguish" for not having understood better while she was alive what disabled people experience. My tears are also my deep regrets; I have in the past been among those who stared. I also get a sick feeling in the pit of my stomach remembering that as a child spending summers in Bradley Beach, New Jersey, I joined the other kids in ostracizing a girl our age who had lost an arm in an auto accident. She always wore a little "sleeve" on her bathing suit to cover her stump. These hurting memories strengthen my zeal to educate us able-bodied about the disabled.

Today, even though we understand a lot more about the disabled, many of us still feel that tiny knot of anxiety when we are around them. We react with superstitious fear to the thought that There, but for the grace of God, go I. If a dreadful accident or disease can befall this person, what's to prevent the same thing from happening to me? Because our lives are not in contact with the disabled, we don't know how to act with them. We are not sure whether to offer a helping hand to a blind person crossing the street or, if we do make such an offer and it is accepted, whether the helping hand should be on the blind person's arm, behind him, or whether we should walk with the blind person's arm tucked under our arm, leading him? Will our gesture be resented as "patronizing" if we offer to help a person on crutches or in a wheelchair? We find it disorienting not to be able to engage in that familiar social ritual—extending a handshake. The disabled often are not able to give us their hands in return.

We also find it esthetically upsetting to be around those whose faces are twisted, whose mouths may drool, whose tongues work aimlessly in their open mouths, whose clothing may be askew or garish with badly chosen colors and patterns. In our hierarchy of esthetic sensibilities, deformities in the lower part of the body, such as those caused by polio, are more acceptable than deformities of the face, such as those seen in the twisted features of some victims of cerebral palsy.[5] Our esthetic sensibilities have been given legal credence. Some communities still retain old statutes known as "the ugly laws" although they are rarely invoked—"No one diseased, maimed, mutilated or in any way deformed so as to be an unsightly or disgusting object" may step out in public. These laws were passed partly to protect pregnant women; superstition said that seeing someone deformed might cause a woman to give birth to a monster. Chicago repealed its ugly law just a few years ago; Columbus and Omaha still have theirs.

Our uneasiness with the disabled results from our lack of familiarity with them. When we see them regularly, we undergo the phenomenon of "the vanishing disability." The 1977 White House Conference on the Handicapped (many official agencies still use the word "Handicapped" in their titles although more are switching to "Disabled") reported that "the more frequent the contact the more positive the attitude. Close contact often leads to perception of similarities, which in turn can lead to liking."

I was friends for years with Frank Scully, a writer and a columnist for *Variety*, the show-business paper. As a young adult Frank had had a leg amputated below the knee and he walked on crutches. During the years of our friendship until his death he stayed frequently in my home on trips to Los Angeles from his home in Palm Springs. I cannot remember which of Frank's legs was the missing one, what his crutches looked like, whether he had his empty trouser leg pinned up behind his stump or sewn at the knee. All I remember is his forceful, loving, humorous personality, our joyous laughter over the breakfast table, and his beautiful leonine head covered with white hair. In my memory, his disability has utterly vanished.

Artist Toulouse-Lautrec as a child suffered a bone ailment that prevented his legs from growing to normal size. He remained a dwarf all his life. As an adult he entertained lavishly, socialized in cafés, and had love affairs with many beautiful women. Startled at first acquaintance by his gnomelike stature, friends would soon say: "The more you see of Toulouse-Lautrec, the taller he grows."

The disabled want all of society to undergo "the vanishing disability." They are angry about governmental policies that keep them in "separate-but-equal" facilities in which many live in buildings designed exclusively for the disabled or work in "sheltered" workshops designed to rehabilitate them by helping them to earn a living but which actually suppress their opportunities to lead lives as normal as possible by keeping them only with other disabled, away from the rest of us.

The disabled want independence to do for themselves the tasks of dressing, grooming, feeding, toileting, working, if society will only provide the tools and facilities for them to do so. They say, "Some part of every physically disabled person is physically abled. We want to get jobs and earn our own living. We need not be helpless if you provide the equipment and the attitudes to help make us self-sufficient. We want you to relate to us as people, not as paraphernalia."

Our psychological approach toward the disabled is the same as that toward the sick—they are "patients" who must be taken care of and who must cooperate with us by getting "well." But they are not "sick," and they are not going to get "well." Their disabilities are going to remain all of their lives. What the disabled want is not "sheltering" but "mainstreaming" into the currents of human affairs. They would like to be enabled to ride in the same buses, to eat in the same restaurants, to work in the same offices, to learn in the same classrooms, to go to the same movies, plays, concerts and art galleries, and to travel as tourists.

When we are in contact with disabled people we quickly learn how they would like us to help them and touch them, if at all, and under what circumstances they will tolerate our curiosity about what caused their conditions. They look upon help as being of three types: *necessary*, *harmful*, and *humiliating*.

This experience of an Omaha man provides a good example of *necessary* help. Showing a television crew the route he would take to go to a job if buses were equipped with lifts for wheelchairs, he re-created an earlier experience in which he was wheeling himself along the street and he dropped a manila envelope from his lap. Having no piece of furniture or railing on which to lean so he could propel himself out of his chair and then back into it, he was helpless to pick up that envelope from the ground. He had to sit there until a pedestrian came by and he could ask, "Would you please pick that up and hand it to me?" In that situation he had no choice but to ask for necessary help.

You are giving necessary help if you hold a heavy door open for a person in a wheelchair. It becomes *harmful* help if you insist on pushing the chair through the door. The person *cannot* hold the door himself; he *can* wheel the chair through the door himself. With all good intentions you may be offering harmful help if you take the arm of a blind woman walking with a red-tipped white cane and propel her across the street while the traffic light is green. Your sudden touch may cause her to suffer the "startle reflex," mentioned several times earlier in this book. Your help could be harmful if a blind person were walking with a seeing-eye dog. A blind girl says, "When Amy [her dog] and I are out walking we're almost like a ballet team working together in motion. Or like a couple dancing."[6] To interrupt this harmonious "ballet" by touching the girl to guide her would be about as helpful as leaping up on stage during a ballet and grabbing a leg of the dancer to help her through a pirouette!

The blind may not want your help at all. You can find out by asking, in a quiet, nonstartling voice, "Would you like help?" If the person says yes, then ask, "Do you want to take my arm or do you prefer that I hold your arm?" Usually but not invariably, the blind prefer to take the arm of the person leading them. You may have to take the person's hand gently and place it up under your arm. The blind can sense instantly through their touch on your arm if you hesitate or stop because of traffic conditions.

Humiliating help may come with a flourish—the "helper" is often doing it more for his own psychological needs. He may be showing off—"See what a nice guy I am." He may offer help with pitying looks, in a loud voice, or with a patronizing "you poor helpless thing" manner. He may barge in and grab the wheelchair and steer it over the curb when the disabled person would prefer to do that himself. A man says about humiliating help, "Don't give me any 'March of Dimes smile' as if I were 'the poster child of the month'."

An easy rule of thumb is that most disabled people want to do as much as possible for and by themselves. Even if they are straining at a task, they may prefer to accomplish it without help. By being overeager to help we may deprive them of a skill they want to master. Their concern is: How much sensation do I have in my body, and where? How much movement do I have in my arms, elbows, wrists, fingers, palms, legs, feet, toes, knees? What tasks I can accomplish without help? A young woman, lying in her hospital bed after surgery following an accident that left her paralyzed, tells of a poignant moment in which she tested whether she could still feel. Laboriously she took a few grains of sugar from a meal tray and put them into her palm. When a fly landed on her hand and she could feel the vibrations of its legs crawling over her skin, she wept with joy. Behind every product with which the disabled come in contact is this question: "Will this increase my mobility, my ability to do for myself?" Mobility means freedom from dependency.

A tool that has great symbolic meaning for the disabled is an ordinary house key with a long metal extender bolted to it, which many require because they don't have the motor skills to insert and turn a key by itself. A key of their own means they are wheeling themselves and hobbling on crutches out of family homes and out of institutions, into homes of their own where many are living alone or sharing with other disabled persons. The Center for Independent Living in Berkeley, which sparked what has come to be known as *The Wheelchair Rebellion*, is a world-famous halfway house where disabled come to live and train themselves in self-sufficiency before going on their own.

As more of them have their own homes and privacy, they are beginning, at long last, to learn to enjoy their sexuality. Our consciousness that the disabled are as needy and desirous of sexual pleasure as the rest of us are was heightened by a sensitive love scene in the Jane Fonda movie *Coming Home*. About to make love for the first time with her boy friend, a veteran paralyzed from the waist down, she asks him tenderly, "Show me where to touch you. I don't know where you have feelings." Even though some may not have the pelvic mobility and genital capacity for intercourse they can achieve the ecstasy of orgasm through oral sex and masturbation. A bill of sexual rights, published by the Sex Information and Education Council of the United States, and meant to educate the public as well as the disabled themselves, says:

1. Urinary incontinence does not mean genital incompetence.
2. Absence of sensation does not mean absence of feelings.
3. The presence of deformities does not mean the absence of desire.
4. Inability to move does not mean inability to please.
5. Loss of genitals does not mean loss of sexuality.

The more we interact with the disabled the more we become aware of their likenesses to us rather than their differentnesses from us. They breathe, as we do. They have skin, as we do. They may sometimes long to be touched and held, as most of us do. They may sometimes prefer to be left alone and untouched, as some of us do. Even if they can't return a touch or squeeze or caress, they long to receive those sensations from us.

Even though the disabled do not want pity, we may nevertheless sometimes be moved to tears. Comedian Jack Benny was walking through the Broadmoor Hotel in Colorado Springs when a woman asked him a favor. Her eighteen-year-old daughter had been blinded in an automobile accident a month earlier and she wanted to meet the famous Jack Benny. Jack walked up to the young woman who asked if she could touch his face. She ran her hands gently around his face and head, then said, "Thank you, Mr. Benny. You've made me so happy." Jack broke down and cried.[7]

Dealing with the disabled, we will like some and not others. We will want to help some and not others, to avoid some and seek out the company of others. We'll touch them on the same basis as we touch other friends, acquaintances or strangers. Laughing merrily with a disabled friend, we may reach over and give her arm a squeeze of joy because we are sharing fun. Saying goodbye, we may hug her even if she can't hug back. Full of good spirits at a party, we may pick up a friend who doesn't weigh much and dance her around in our arms, or take her arm and twirl her in her wheelchair, or hold on to the chair and dance it around as if the chair were our partner. Most of the disabled are not fragile; they enjoy fun and motion as much as the rest of us do.

A major adjustment we do have to make, as I did, is in being patient. It takes more time to deal with them, because of their necessarily slower pace. Walking beside a person on crutches or in a wheelchair, we have to slow our pace to accommodate theirs (although I have seen people on crutches and in motorized wheelchairs move at pretty fast clips). The disabled themselves feel impatience with their pace. Bill Whiting, an artist who paints holding a brush in his mouth, says on *The Turning Point*, a television series by and about the disabled (the host, Tom Campbell, is blind), "I am learning since my accident that I have to be patient and it's hard."

Some day we will find ourselves on a bus equipped with special steps and wide doors to lift wheelchairs from ground level into the bus. We may feel impatient about this enforced slow-down, waiting for the chair to be lifted in. But the few moments this requires could provide some healthy calm for us in our speeded-up, frenetic culture, giving our hearts and minds a few seconds for contemplation and admiration. Rather than fretting and becoming impatient, I intend to devote that time to marveling at the indomitability and stamina of the human physique and spirit.

Most of us, being people of goodwill, will relate well to the disabled. Touching them, we'll learn that a muscle that doesn't work is not necessarily in pain. We can squeeze, clasp, tickle, lean on, caress a disabled man, woman or child exactly the way we might an able-bodied person. (Remember the charming scene in the movie *Coming Home*, in which Jane Fonda sits on Jon Voight's lap and they go sailing along in his motorized wheelchair?) *Muscles atrophy, skin cells do not.* The feel of warm flesh can help the disabled cope better with the cold metal and plastic of wheelchairs, crutches, braces and other equipment. It can make them feel accepted and cared for. Ultimately, flesh is more powerful than metal.

10 Touching in Sports

Imagine this scene on a balmy day centuries ago. On an open green, men, women and children are running about kicking at an object, perhaps a smooth stone or a piece of cloth woven into a ball. Each one tries to keep the others from getting to the object. A child wraps her arms around her father's waist, laughing as she tries to hold him back and he playfully pulls out of her grasp. A woman, racing past a toddling boy, sweeps him up into her arms and continues to run with him; he shrieks with delight at the sensation of being lifted and carried. Adolescent boys and girls smile at the pleasure they feel when their shoulders and hips make fleeting contact as they bump into one another. The frolicking players fill the air with lovely sounds of merriment. Not having any mission to score points, they do not keep track of whose foot makes contact with the object the greatest number of times nor how far he or she propels it with a kick. Victory is not their goal. After a while, exhausted and out of breath, they flop on the ground, happily drawing renewal and nurturance from the earth's warm caress. Contentedly, some of them reach arms toward one another, fingers interlocking, acknowledging the pleasure they are feeling.

Our early ancestors were experiencing play as Nature intended it—pleasurable movement that exercises our muscles, stimulates our heartbeat, increases our intake of oxygen—helping us to lift, however briefly, the "gravity of anxiety" off our psyches and to rid our bodies of the toxic effects of stress. In giving play, her antistress formula, to the human race, Nature carries on her relentless drive for the survival of the species.

Would our primitive relatives be interested in playing if they were hungry? Of course not. Before they could fill their need for fun, they had to fill their need for food. The hunter, fleet of foot, dexterous of hand, went foraging with a weapon, a roughhewn spear, chasing a prey to slaughter. The hunter had to win out over his prey; otherwise, he and his family would perish. Nature, again carrying on the survival of the species, made us hunters as well as players.

Along with our drives to hunt and to play, we have been given a drive to "bang into people."[1] Everywhere in the world people play touch games such as "tag," a simple way to bang into one another. Children do not have to be taught the game of tag, they do it instinctively, touching someone and then running away playfully. Archaeological records show that the human race has always played variations of tag. This drive represents yet another of Nature's survival techniques; she intends for us to make contact through touch.

These drives that are built into our genes—to hunt, to play, to touch—are the basis of the human race's preoccupation with sports. No matter what a nation's politics, economic system, technology, climate, terrain or size, its citizens avidly play games as participants and frenziedly cheer games as spectators. Games played around the world number in the thousands. Despite variations in different lands—the Brazilians and Chinese prefer soccer, the British prefer soccer and rugby, Americans prefer baseball and football—the world's games have more similarities than differences as we all carry on ancient patterns of behavior originating in our biology. All of us, at times banging into each other, are sometimes players, sometimes hunters.

According to Webster's New Twentieth Century Dictionary, Second Edition, a *sport* is "any activity or experience that gives enjoyment or recreation"; or "such an activity requiring vigorous bodily exertion." *Sport* is a contraction of the word *disport*, which originally meant "to carry away from work." Sedentary games such as cards, chess, mah-jongg and backgammon fit part of this definition—they give enjoyment or recreation—but since they require mental agility rather than physical exertion, we classify such games as pastimes rather than as sports.

As we "disport" ourselves, when are we players and when are we hunters? If we are jogging, surfing, hiking, skiing, swimming, bicycling, lobbing tennis balls, roller-skating, ice-skating, or skateboarding for the major purpose of achieving well-being, exercise and good health, we are players. I saw a perfect example of this as I was walking to the market on a Saturday afternoon. I watched two women roller-skating along Sunset Boulevard. The happy looks on their faces, their musical motions as they pumped their legs and arms and flowed back and forth from side to side across the sidewalk was the epitome of the human spirit enjoying a play sport.

But let us say these women were competing in a roller derby; instead of skating playfully on Sunset Boulevard they were skating around a rink trying to win a prize. Gone would be their happy expressions and carefree manner. Their bodies and faces would probably be contorted with tension as they tried to get ahead of others. Perhaps they would jab their elbows hard against other skaters to impede them. The women, engaged in the same sport, would no longer be players; they would be hunters out for a "kill."

When our primary motive in sports is to win out over, to defeat, a person or a team, we become hunters in pursuit of a prize, a win, a "kill." Our ancestors hunted food for the body, we hunt food for the psyche—glory, money, power, the satisfaction of being best. Professional sports with their "kills" of huge salaries and national fame, are hunter sports. Football, baseball, basketball, hockey are the most common "hunts." The hunter's "weapon" is the bat, the hockey stick, the football or the basketball; his "prey" is the goal post, the basket or home plate. Zoologist Desmond Morris says:

> Viewed biologically, the modern footballer is revealed as a member of a disguised hunting pack. His killing weapon has turned into a harmless football and his prey into a goalmouth. If his aim is accurate and he scores a goal, he enjoys the hunter's triumph of killing his prey.[2]

It is not surprising that spectators at hunter sports shriek at players, "Kill 'em, kill 'em," or that coaches send teams on to the field exhorting, "Go out there and kill 'em." As hunters, we engage in sports primarily to win, which adds to stress; as players, we engage in sports primarily to attain well-being, which lessens stress.

It is the human's drive for sensory stimulation, combined with our drives to play and to hunt, that make sports so satisfying. Sports provide us with our greatest variety of tactile sensations; we touch not only people but also the elements—earth, air, water, the forces of Nature. By doing so we may attain some of life's most euphoric experiences, possibly even enter altered states of consciousness from time to time. Physician Andrew Weil, author of *The Natural Mind*, says that the human brain craves this exalted state as a respite from its constant work.[3] Roger Bannister, the first human on record to run a mile in less than four minutes, tells that while running he would feel himself get lighter and lighter with each touch of a foot upon the ground. It was as if the earth were a giant trampoline bouncing him into the air with every pump of his leg and foot. Sometimes he would lose all sense of his body, feeling weightless, a pure mind, flying through space, unattached to any physical being.

Some joggers have similar euphoric sensations when they reach their "second wind." For many, their touch sense becomes acutely heightened. They can almost feel the weight of a breath of air on their tongues and in their noses. They become aware of the delicate fanning of air floating past their cheeks, of their arms moving back and forth rhythmically massaging their sides. They can feel drops of perspiration trickling down their faces and down their bodies inside their clothing.

Swimmers, divers and surfers may also experience a second wind in which they become aware of the "sensual, slipstream touch of water." For some, the flow of water against their bodies triggers intense elation, almost as if they were having sexual orgasm. Ed Cowan, a novelist and sports lover, writes of the sensuality of surfing:

> In body surfing there is the pleasure of "tapping" a natural energy flow. The athlete catches and rides (Nature's) energy and it feels good. Let's suppose that it's a good surf and the largest swells are running up to twelve feet. As the surfer looks up at the approaching wall of water, the power of the wave gives rise to that little ball of fear at the pit of the stomach. But as he catches the wave and makes that five-to-six-foot slide to its bottom, pulled down by gravity and driven forward by wave action, there is a physical rush that can only be felt, not imagined. The rush continues as the wave breaks about the surfer and drives him onward. This is pure, physical pleasure, a tangible charge through the body imparted by the wave. Additionally, there is a mental pleasure—the satisfaction of knowing that for twenty-odd seconds the surfer is immersed in and riding

one of Nature's wildest beasts. Though he is being tossed about like a rag doll in a washing machine, it feels good. The energy of the wave is a pleasant sensation to the body.

Mastering time, space, water, air and earth in sports, we often feel powerful, at one with Nature. When we are most profoundly fulfilling Nature's intention to renew our bodies and souls in play, we come closest to achieving "the perfect rhythm that exists in each of us, in tune with the silent pulse of the universe."[4]

How enriched our lives could be if all of sports provided such luminous sensory experience! But another aspect of touching in sports, instead of ennobling us, brutalizes and dehumanizes us. This is the malicious "game within a game" in which players punch, kick, gouge, ram, choke or stomp their opponents. This secret game is played below the surface of the public game. "Winners" are those who are adept at perpetrating dirty touches without being caught and fined or otherwise penalized. It is this secret game that often triggers fist fights among players after a game officially ends. Enraged by the dirty touches, a victim rushes to a malefactor and takes a swing at him, with teammates sometimes joining in. (In ancient Greece officials publicly whipped athletes who committed infractions of the rules, which is somewhat like a parent beating a child to teach him not to hit.)

Violent touching is illegal within the rules of a game but it is subliminally sanctioned by players, coaches, team owners and other officials all in the name of "winning." Woody Hayes, then the football coach at Ohio State University, told sports writers, "We teach our boys to spear and gore. We want them to plant their helmet right under a guy's chin. The boy who blocks with his head down gets hurt. I want them to stick that mask right in the opponent's neck."

Social scientists are of two schools of thought about violent body contact in sports. One group believes that it serves a socially useful purpose by helping us to discharge vicariously our own feelings of violence. A hockey player punching an opponent is acting out for us the rage many of us feel at society, government, employers, children, spouses and others. "Boy, give it to 'em, that's what I'd like to be doing to somebody." The other group believes that when we see violence being accepted by the populace, we tend to become more violent ourselves; observing such behavior, particularly by highly regarded sports heroes, gives us permission to do the same. Anthropologist Richard Grey Sipes describes the two beliefs as the "Drive Discharge Model" and the "Culture Pattern Model."[5]

Researchers have done studies validating both schools of belief—that the violence we see in sports (and also in films and on television) may either provide a catharsis for us or instigate violence in us.

My own belief fits in with what Los Angeles psychiatrist Barry Alan Smolev calls the "Frustration Theory." This suggests that other factors contribute to whether or not we imitate violence that we see in sports. If we live in an environment of unmet needs, a personal economy of scarcity, we are more likely to commit violence out of frustration. If we live in an environment of personal abundance with our needs fairly well met, we can see violence but not become violent ourselves.

We may begin to see some lessening of violence in sports as a result of growing numbers of young athletes (including high-school students) being crippled for life, often with spinal injuries, when they lose that game within a game. And also as a result of a precedent-setting lawsuit won by Rudy Tomjanovich, a Houston Rockets basketball player, against Kermit Washington and the Lakers. As this book goes to press a Houston jury has awarded Tomjanovich $3,300,000 because he suffered a broken jaw and nose, a puncture of the brain cavity, a torn tear duct and the loss of teeth from a single devastating punch. The jury said the Lakers were negligent in controlling players "with a tendency for violence."

Counteracting my stress when I see athletes pummeling one another is the warmth I feel when I see teammates, and even opponents, touching each other in admiration and good sportsmanship. Sports lovers are familiar with the "tush touch" and the "palm pat." When a player scores, a teammate is sure to give him an approving pat on the behind, or the scorer runs with a palm held upward to receive his reward, the accolade of an admiring slap into his palm from a teammate. Their two hands are joined in "applauding."

Several memorable scenes of touching in sportsmanship stay in my mind. One occurred during the tryouts of the women's 400-meter run in the 1980 Olympics. Repeatedly, a winner would go over to a loser and throw her arms around her in consolation, as if to say: "You're great, even if I did run faster than you did." Another was the heart-warming sight of contestants in the 1980 Boston Marathon, seconds before the starting gun, reach out to one another with "good luck" handshakes. Still another scene appeared on a Sunday afternoon NBC-TV program, *Sports World.* James Scott, an inmate of Rahway State Prison in New Jersey, won a boxing match with Bunny Johnson. He walked to Johnson, gave him a comforting embrace, pressed his cheek lovingly to Johnson's cheek, and whispered to him. This moving scene was heightened by the fact that Scott, serving a sentence for armed robbery and parole violation, was seeking to become light-heavyweight champion while behind bars. He is officially recognized as a contender by both the World Boxing Council and the World Boxing Association, and must find ranked contenders who will box with him inside prison walls. Johnson had been willing to do so.

Rituals of touch, especially those motivated by superstition, are common in sports. Boxers traditionally touch gloves before a match, swordsmen touch foils, hockey players pile their glove-clad hands one on top of the other and go through a chant; they also skate in front of their goal and tap the goalie on the pads for good luck. John Wooden, UCLA's noted basketball coach, now retired, would wink at his wife, tug at his socks, pat the knee of his assistant coach and tap the floor. A superstition prohibiting touching maintains that players who are having a losing streak should not touch players who are about to compete; they might give negative vibrations.

A powerful and controversial taboo against touching in sports is directed against the most intimate of all touching—the sex act. Many coaches forbid players to have sexual relations prior to a game; they believe that the energy used for copulating will diminish strength for playing. This stems from an ancient notion of "limited good," which assumed that a man has just so much energy and so many orgasms in him, a quota for his lifetime; each time he has sex he depletes that precious quota. Some coaches make a bed check the night before a game to make sure that players have not slipped out. One college coach insisted that his married players move into a dormitory several days before a game so they wouldn't be near their wives and succumb to temptation.

Ancient beliefs die hard. They were formulated before we had knowledge we now possess, that the release of sexual tension, especially for virile young athletes, is more likely to energize than to exhaust them. For several days preceding a Super Bowl game, the Minnesota Vikings were kept away from their wives and girl friends; the Pittsburgh Steelers were not. The Steelers won, by a score of 16 to 6.

There is some logic to the idea that what may *accompany* the sex act—partying, drinking, staying up late—can detract from a player's stamina, but team rules are confusing sex with sociability. If players have sex and don't stay up late socializing they can perform as well as, and probably better than they would without sex. Joe Namath told *Time* magazine of doing both. "I went out and grabbed this girl and brought her back to the hotel and we had a good time the whole night. It's good for you. It loosens you up for the game."[6]

Athletes rail against the no-sex restriction and ignore it. Julius Erving reveals that he conceived one of his children a few hours before a game, and played well. Sex researcher William Masters notes, in *Family Health* magazine, "Restricting a player from sexual relations before a game is hogwash. Pregame warmups use more energy than sexual relations."[7]

This taboo will go by the boards as men and women increasingly play and make body contact on the same teams. Many school teams have both boys and girls playing hockey, football, soccer, softball. Women and men are competing in shooting and equestrian events, in horse racing, auto racing, tennis. The time is not far off when unisex sports will be common, at both amateur and professional levels. Some day soon a pair of male and female lovers will be playing on the same team. They will have sex every time the mood and the available free time coincide. The ecstasy and relaxation they derive from their physical closeness will so invigorate them that they will achieve high scores, carrying their team to victory. Their coach and the sports world will then hail lovemaking as a great body-contact activity.

Heterosexual lovemaking, that is. Anxious to maintain its supermasculine image, the sports world is sensitive to intimations of homosexuality. In the fall of 1975, Lynn Rosellini, a *Washington Star* reporter, wrote nationally syndicated articles claiming that many professional athletes were either gay or bisexual. "Sports fans gasped when they read these articles. Irate letters denouncing Ms. Rosellini inundated newspapers. How could she imply that any red-blooded American boy could be 'queer'?"

Then, in November 1975, Dave Kopay, former running back for the Washington Redskins, admitted that he was gay. Furthermore, he claimed, there were others like him who were afraid to admit their sexual identity for fear they would become the objects of hate campaigns. Kopay later published a book in which he said:

> On the field we can get away with all kinds of physical affection men wouldn't risk showing anywhere else. We aren't ashamed to reach out and hug. After a touchdown you see men embracing on the field like heterosexual lovers in the movies. We were able to hold hands in the huddle and to pat each other on the ass. I think these are healthy expressions of affection. What is unhealthy, I think, is that we are so afraid of expressing ourselves in the same way outside of the stadium, out of uniform.[8]

Sports psychologists all make the same point: In part, male fans love team sports so rabidly because such sports act out suppressed desires to be close to other males. The game with the largest number of body contacts is the game with the largest fan following—football—in which players block, tackle and huddle. Anthropologist William Arens, commenting on the psychological protection that footballers get from their heavy garb, says, "Dressed in this manner, the players can engage in handholding, hugging and bottom-patting, which would be disapproved of in any other context but which is accepted on the gridiron without a second thought."[9]

"Grab-ass" is common locker-room horseplay when a team wins. Exuberant naked males may acceptably grab a teammate's buttock. If a player is tempted to touch another's bare skin but is afraid of being thought gay, he indulges in another common locker room horseplay—the "towel flick"—an accepted surrogate touch. Football has been called "America's Number One fake-masculinity ritual," displaying the very behavior that homophobic athletes and spectators profess to despise.[10]

Several years after the revelations of homosexuality, people in sports are still sensitive on that subject. Jim Hill, a Los Angeles sportscaster, asked sports figures if they thought that touching in sports helped men to express vicariously what they were afraid to express off the field. Hill juxtaposed shots of anthropologist Alan Dundes saying yes, and shots of athletes and coaches saying no, with close-ups of the hands of the Los Angeles Rams and New Orleans Saints football teams. "That idea is stupid," snarled a sports figure while the camera zeroed in on athletes holding hands as they ran across the field, hugging with full frontal hugs, and kissing on the cheeks and on face masks.[11]

Some day an enlightened human race will make it acceptable for men to show affection unashamedly. When men no longer require sports to experience vicariously physical closeness with each other, will our interest in sports then decline? Not at all. Sports are a glorious adventure allowing us to test our muscles, bones and sinews, and to put our bodies through shapes, forms and motions that give us intense esthetic pleasure. A basketball player's gazellelike grace as he leaps upward and tosses the ball through the hoop, the rhythmic ripple of a batter's arms, shoulders and torso as he hits the ball are beautiful sights. I get a thrill observing the deftness of the human body when a football player whirls suddenly in a surprise maneuver and flips the ball into another's hands. I laugh with pleasure when I see a "winners ballet" in which a team hugs en masse and jumps up and down with joy, or in another sports "dance," one member leaps into the arms of another, straddling his body with his legs while the two prance around exultantly. Can you picture corporation executives, having just closed a million-dollar deal, or surgeons, having just completed a delicate operation, expressing joy the same way? They may feel like doing so, but such behavior is acceptable only in sports. Nowhere but in sports could you see burly, brawny

men rejoice at a victory by publicly kissing each other full on the lips, as Russian hockey players do.

Over and over, sports feed our eyes and souls with feats of agility and suppleness. Listen to the screams as baseball player Willie Mays, seemingly defying gravity, leaps high into the air and catches a ball with one hand; hear the thunderous applause as a ski jumper floats several hundred feet through the air with his body almost horizontal; and know that the human race will forever thrill to sports. We will always be what Dutch philosopher Johan Huizinga calls *Homo Ludens*—"Playing Human"—celebrating the play element in the human spirit.

There *is* one factor that could conceivably change how we feel about some sports. That is the ultimate capacity of the human body to compete against time and distance. A time will come when the body is not capable of creating any more records of strength, speed or endurance. No runner anywhere on earth will be capable of running faster than many others are doing. No swimmer will be able to swim faster or farther than others are already doing. Sports historian and sociologist Allan Guttmann asks, "What will happen to our obsessive quest for records when athletes finally begin to reach, as eventually they must, the limits of human possibility?"[12]

When the body reaches this ultimate capacity, fewer of us may want to attend events, such as swimming and track meets, which are for setting records, while our interest continues in football, basketball, baseball, hockey, figure-skating, tennis—those sports in which the drama of the unexpected can happen or at which we can admire brilliant choreography, maneuvering and outsmarting.

Reaching the limits of human possibility could have another result. Many of us would lose a satisfaction that sports now provide to us—vicarious pleasure from others' accomplishments. Hearing of new records, we take pride in being part of the human race, one of whose members has performed so magnificently. We live taller. Should we be deprived of this pleasure from engaging in sports as observers, more of us may become involved in sports as participants so that we can experience, at first hand, what bodies can do.

Stimulated by increasing leisure time and our growing awareness that we must exercise as part of health care, we are becoming a nation of sports participants. More than a hundred million Americans of all ages are joggers, tennis and racketball players, roller-skaters, skateboarders, bicyclists, surfers, skiers, gymnasts, bowlers, swimmers. Sports are no longer activities in which we engage only while we are in school and then drop. Watching the artistry of Olga Korbut, a teen-age Soviet gymnast, in the 1972 and 1976 Olympics, many men and women, including me, were inspired to recapture the joys we had known as schoolchildren working out on parallel bars and gym horses. One night at the Hollywood YMCA, the years slipped away for me as I pressed my forearms down on parallel bars and swung my legs back and forth, recalling delicious sensations of floating and weightlessness that I had not felt since Weequahic High School in Newark more than forty years before. I was delighted by my geriatric agility as I hoisted my body above the bars and discovered that I could still swing astride them. I was klutzy, but a klutz who enjoyed herself.

Some observers believe that this upsurge in physical activity is partly a reaction to our being saturated by televised sports. A nation of sports participants could become a boon for "sports widows," women who are angry because their husbands or partners sit indoors passively in front of television sets on glorious Sunday afternoons. A man learning to enjoy the release of bodily tensions and the feelings of well-being he can get from participating in sports might say, "I'm too restless to sit. Let's get out and go roller-skating." Or bicycling. Or jogging. Or even walking. As our bodies become conditioned to exercise, we find it harder just to sit. I am so accustomed to expending energy in short-distance jogging that when I am housebound, I feel as if I could jump out of my skin. Once I got up from a sick bed and loped around the neighborhood. My lover paced me in his car, just in case a policeman might become suspicious at the sight of a woman in nightgown, robe and track shoes running after midnight around West Hollywood! I am convinced that the pleasurable exertion, which my body craved, speeded my recovery.

There will not be any lessening of our interest in sports, nor will there be any decrease in their commercialized aspects. Sports are big business. Sales of tickets and television rights to games, money spent by teams on travel, equipment, advertising, and by fans supporting their favorite teams, all contribute to the national economy. It was an eye-opener for me to be on a plane once with many USC alumni and to learn that they structured their social lives traveling as "groupies" to cheer their alma mater's teams, spending lavishly on hotels, meals and transportation. (The word *fan* is a shortened version of *fanatic*. Early baseball teams called their fans "cranks," probably because fans were "cranky" when teams lost—no different from today.)

By paying huge salaries, commercialized sports also help to keep alive the American dream that it is still possible to become an instant millionaire. O.J. Simpson, Bruce Jenner and Magic Johnson are only three of many champions who have kept that dream shining by becoming millionaires almost overnight through sports earnings, television commercials and movies, and by endorsing products.

By attracting huge audiences, sports fill a deeper need for us. Theologian Harvey Cox says we need in our souls to experience "festival, pageantry, spectacle."[13] Early societies filled this need with agricultural festivals and religious pageants. Today, sports provide one of our few opportunities to be part of spectacles—as do rock concerts. It is thrilling to walk into an arena and see thousands of people in festive mood. Our blood tingles as we see vivid clothing, hear boisterous shouts of revelry, anticipate the excitement to come. "High" on feelings of joy, many of us dance our way to our seats to the rhythm of blaring music. Spectacles allow us to behave publicly in ways that we are not otherwise permitted to do. We can scream, yell, shriek, sing, clap, wave flags and pennants, blow horns, wear bizarre hats, exchange opinions with strangers, and hug and jump up and down expressing our elation when a team scores. Thousands of people who were strangers to each other did exactly that, weeping with joy clasped in each other's arms, when the American hockey team defeated the Russian team in a surprise victory at the 1980 Olympics. Such events are great social levelers in which "economic status and educational differences are erased when people gather together in the same stands in what often becomes a massive shedding of inhibitions, as fans rub shoulders high and low."

In some socialist countries this need for spectacle is filled not only by commercialized games but also by meets in which hundreds of thousands participate simultaneously in gymnastics, *tai chi chuan* or calisthenics, sometimes in formations spelling out political slogans. Friends who have attended *Spartakiades*, as they are called in the Soviet Union, Czechoslovakia and the Democratic Republic of Germany, or *Yun Tung Huei*, as they are called in the People's Republic of China, tell me these events with masses of humanity moving in unison, provided the most spectacular sight of their lives.

In addition to maintaining our interest in commercialized and competitive sports, we will place increased emphasis on the New Sports—noncommercial, noncompetitive events in which men, women and children play together. No one will keep score. The goal will be to enjoy running about exerting our bodies, making pleasurable contact banging into one another without the malice and pain of "the game within the game" and to encourage good feelings among people who are having fun together. The Pueblo Indians of New Mexico have a tradition which sums up the philosophy of the New Sports. Their games and races often involve old men and young boys in the same event. The purpose of the sport is "not to beat someone else but only to do one's very best."

That it is possible to have fun without adding up a score is shown by a game played by teams from various embassies on the grounds of the American Embassy in Moscow during winter months. "Broomball" is played on ice skates, like hockey, but the sticks are brooms and the puck is a plastic disk. When a television newsman filmed a game between employees of the American and West Germany embassies and asked what the score was, a player replied, laughing, "Oh, we don't care about that, we're all just enjoying the exercise after sitting at our desks so much."

In Japan a game similar to football, *kemari*, has neither winners nor losers. The ball, made from deerskin, is kicked from one person to the next who has to kick it into the air before it hits the ground. The players' skill is in keeping the ball in the air, competing against time and gravity, not against one another.

An aspect of competitive sports that disturbs many psychologists and sociologists is the pressure we are putting on children to be killer-hunters in such games as Little League baseball. Many parents have loving intentions when they encourage children to play such hunt sports. They want their children to have fun while they are exercising, but the opposite occurs. Go to a park some Saturday morning and study children's faces as they play. They look worried, tense, despairing, disappointed in themselves if they are losing. Even winners, as they undergo "threat of defeat and anxiety about the outcome"[14] are too tense to be feeling pleasure. Except at the exciting moment when someone is scoring, you hardly see children laughing or smiling at their "play."

Sports psychologists are encouraging kids to take up "player" sports rather than "hunter" sports. They also recommend "no trophies, medals or awards for sport victories for children" and that parents and teachers emphasize self-improvement and friendship rather than competition in sports skills. As the Chinese say, "Winning or losing is temporary; friendship is everlasting."

When Shirley MacLaine toured China with a group of American women, they watched children playing a game of tug of war. Some of the women started choosing sides and shouting encouragement. The children stopped in the middle of the pulling and chanted to each other, "Friendship first! We learn from you. You learn from us. We learn from each other!"[15]

What I am learning as I write this chapter inspires me to change what I say to my older grandson when I happen to see him after he plays a Little League game. Conditioned as most of us are by the win-or-lose, killer-hunter ethic, I have been asking, "How did your team do? Did you win?" Now I say instead, "Did you enjoy playing with your friends?" preparing him for the New Sports, which emphasize bodily pleasures and companionship. This philosophy of sports is being promoted by The New Games Foundation (P.O. Box 7901, San Francisco, Calif. 94120), a group started by Stewart Brand, onetime publisher of *The Whole Earth Catalog*, which encouraged the back-to-earth-and-nature movement of the 1960s. The Foundation teaches people to organize "play communities" in which everyone, regardless of age or size, can join. It seeks to replace traditional high-pressure win-or-lose sports with cooperative sports that make everyone happy. Its motto: "Play hard, play fair, nobody hurt."

Playing a typical new game, Planet Pass, you would find yourself in a group of men, women and children lying on their backs in two lines with their heads all toward the center. You all raise your hands as an Earthball, a rubber-and-canvas globe six feet in diameter painted with continents and oceans, starts passing down the line. As soon as you have passed the ball along, you run to the end of the line (which may be quite long, with many participants) and quickly lie down, ready to receive the ball again. People shriek with laughter as they constantly run and position themselves to see how long all of them can keep the Earth aloft in their cooperative play.

In the Lap Game, people stand in a circle, shoulder to shoulder, and then turn to the right. Very gently, everybody sits down simultaneously on the lap of the person behind him. New Games Festivals have had as many as 1,306 people sitting on one another's lap in Auckland, New Zealand, and 1,468 in Palos Verdes, California, hilariously enjoying the camaraderie and the joy of people supporting people as they try to balance themselves and keep their circle from falling apart.[16]

I've played Stand Up, in which two or more people sit on the ground, back to back, knees bent and elbows linked. You all attempt to stand up together. This buttock-to-buttock and back-to-back contact brings many muscles into play; I found this game surprisingly energetic. It was fun having this socially acceptable way of "banging" behinds with men and with other women. The exhilarating massage did more for my psyche than any winning score could have done.

A noncompetitive ethic makes it easy for family members of all ages and sizes to play together. No younger child need be told, "You can't play with us, you're too little." Family play also provides a fine milieu in which those who are self-conscious about touching can become comfortable doing so as they heap, huddle or pat together. A father told me this with much emotion in his voice:

> My sons and I were inspired by a television report of an annual event in Newport Beach where people build huge sand castles on the beach. So we went out to the beach armed with rulers, wooden blocks, and other equipment and we spent the day building a Sand City. We were working away, patting the wet sand, when I suddenly realized my

> hands were touching their hands. This was the first time in many years I remembered touching my sons' hands. I felt like sweeping my kids up into my arms and hugging them, but I was afraid I'd embarrass teen-agers doing that. I'll never forget the joy I felt. Everything seemed to take on a glow. If I could be that close to my sons, the world was suddenly all right for me.

A common activity that opens up vast possibilities for family fun, joy and physical closeness, but has not been classified as *sport*, is dancing, which fits very well the definition of sport. Dancing gives enjoyment, requires vigorous bodily exertion, carries us away from work, and recreates our energies anew. All family members can join in disco and rock-and-roll dances, improvising their own free-spirited choreography. Family twosomes can do ballroom dancing. At a fund-raising dinner Ava Helen Pauling, then in her seventies, wife of scientist Linus Pauling, jitterbugged with a teen-age grandson. People applauded, as much to express their pleasure at the warm feelings of kinship the dancers generated as for their skill and grace.

A favorite "sport" of mine is folk dancing. Among its joys is the sight at folk-dance festivals of three generations, parents, children, grandparents, dancing hand-in-hand or arm-in-arm. I get misty-eyed when I see a young couple holding an infant in their arms between them so they can introduce their child to the merry rhythms of polka-ing, jigging, hopping, twirling, leaping—a family hugging in the sport of folk dancing. On the theory that the family who enjoys playing together will enjoy staying together, family-life specialists are urging families to use sports to build family stability.

Philosopher Johan Huizinga says that the drive to experience pleasure through play is at the core of our existence. The Judaeo-Christian code, which lauded work and decried play as idleness, has prevented us from according play its full measure of importance in our lives. We work to buy ourselves the security and freedom to play. All work is a drive toward play.

A goal would be for people to enjoy their work so much that it feels like play. We can begin to achieve this by bringing a psychology of play into the workplace. "Why choose to regard play and pleasure as 'luxuries' to be tolerated only as a reward for drudgery, rather than as *states of mind* that can be cultivated in the course of any occupation?" asks Lane Jennings of The World Future Society. He suggests that "Instead of jogging for an hour in the park, the boss might jog along Wall Street after work hand-delivering mail that would otherwise have been tied up for hours or days by the overloaded postal service." A desk worker might get exercise during the day by helping to load crates onto delivery trucks, enjoying using his muscles, handling items, products, packages with which his hands are not in touch in the course of his work, expanding his sensory knowledge and perceptions. We could also introduce the workplace into sports. "Golfers might add a sharp-pointed stick to the clubs in their bags and agree to subtract one stroke from their final score for each piece of litter they pick up going around the course," says Jennings. He calls these "productive recreations"—the productivity of work that provides us with the pleasure of sports.

I now look upon some tasks as productive recreations, and I am enjoying them more. Sweeping leaves, I imagine myself playing a solitary game of broomball, working up a sweat as I huff and puff. Washing dishes, I choreograph an imaginary dance swishing rhythmically from soapy water to clear water. At this moment I am enjoying the powerful feel of my fingertips on the typewriter keys as I "play" at typing these words. Recently, I helped friends to move by carrying cartons from their apartment to an elevator. I felt great, testing my muscles to see how many cartons and what weights I could carry, stretching my arms to learn the largest girth I could manage. People kept saying, "Don't work so hard, Helen, sit down."

They didn't understand. I was not *working* hard, I was *playing* hard—at my one-person "sport" of carton-carrying.

All of us are going to be more involved in sports—at home, at work, at play. More of us are becoming sports addicts. A psychiatrist says of his addiction to jogging: "I can't *not* run." Designers of industrial buildings are including gymnasiums, so that employees can exercise during lunch breaks, during rest periods, and after work; many companies already have swimming pools. The American Alliance for Health, Physical Education and Recreation recommends that more men and women train for careers as sports psychologists and sociologists in preparation for the nation's physically active life styles ahead.

A school in San Diego is an admirable model of how sports can help to solve problems of truancy, delinquency and integration. Russ Batza is the principal of Fulton Elementary, a so-called "magnet" school, which offers enriched programs to encourage black and white parents to send their children there. Students have an extra hour of sports every day—volley ball, tennis, soccer, softball, basketball, gymnastics, swimming, track and field, and aerobic exercises. Batza told me that athletics have ended most tensions and truancy. "Kids are eager to come to school so they can participate in sports."

A California physician believes physical activity to be so important that he will not permit his children to watch television until they have exercised first. James Holmes of Scotts Valley has rigged up a "Pedal-a-Program Exerciser." Until someone rides the bicycling device to build up power, the set can't be turned on. One hour of pedaling earns one hour of power. Author James Michener, saying he owes his life to sports, credits basketball with preventing him from becoming a juvenile delinquent and with getting him into college on a scholarship. Later, it was vigorous physical activity that kept him fit after a heart attack. "Sports," he said, "are becoming a major force in American life."

This "major force" provides us with many built-in opportunities for touching—making body contact as we play games, group-hugging to celebrate a victory, embracing shoulders and clasping hands to show sportsmanship, patting each other out of sheer exuberance. Sports free us to use our sense of touch as Nature intended us to do.

As more of us learn how good we feel when we make pleasurable body contact, we may become less willing to condone in the name of "play" the vicious body contact of smashing a knee into a groin, stomping a cleated shoe on an opponent's foot, slamming a face mask down onto a neck (a common cause of spinal injuries). ("Professional football," says sportscaster Frank Gifford, "is like nuclear warfare. There are no winners, only survivors.") We would want sports to add to the world's pleasures, not its pains.

11

Touching in Schools

If you are reading this when schools are in session, this scene of physical touching in a school is probably occurring at this moment, or will happen on the next school day. A child, eight, nine, ten years old, male—it happens to boys more than to girls—is standing before two adults, one his teacher, the other the principal or another authority figure. The child is putting up a show of bravado, acting as if "I don't care what you do to me, I can take it," while his heart pounds with anxiety. At the words "Bend over" the child leans over a desk or a chair while one of the adults hits his behind with a wooden paddle, one, two, three, four, perhaps many more times—depending upon how angry the beater is at the student, her job, her own children, her husband, her whole life. With each blow, the teacher may be exorcising her own pain and rage at frustrations unrelated to the student.

As he feels each *whack!-whack!-whack!-whack!*, the child's body is prepared for "fight or flight"—an action that he may not take. Instead of being filled with obedience and respect, he feels rising hatred and rage, a desire for revenge against the grownups who are subjecting him to this pain and humiliation. To get the feeling of what this must be like for a child, I took a wooden cheese-cutting board shaped like a paddle from my kitchen and whacked myself across the butt several times. It *hurt.* I well believe that a student would be thinking: "Boy, am I going to make you pay for this!" Teachers and administrators who disapprove of corporal punishment believe that many acts of arson and vandalism in schools are a result of revenge by students who are beaten by school authorities.

When the child and the teacher return to the classroom, the "ripple effect" is evident in the tense faces of the other children. Some of them may jeer at their classmate later, during recess, partly out of relief, but they are all scared, uneasy that it could happen to them. Underlying their anxiety may be an even deeper concern. To children, teachers and principals represent the world of "officialdom"—institutions run by government. If government permits more powerful grownups to beat less powerful children, where can children turn for protection? Government becomes something to be feared and hated, not respected. Corporal punishment of a peer often results in an entire class being pitted against the teacher.

"From personal memories as a small child in a Nebraska one-room country school," writes Barbara Fox, cofounder of Houses of Hope of Nebraska, juvenile-care centers, "I can testify to the psychological pain and damage that are inflicted on those who must witness corporal punishment. How well I remember the confusion, sadness, dread and anger I felt when, for some unknown reason, Lucille, a tiny second-grader, was taken to the coatroom and spanked hard almost every day. Later, in a small-town grammar and high school, the 'big boys' who smoked behind the barn were whipped with a rubber hose by the principal whom we feared and despised."

Such scenes may occur in parts of forty-six states at this writing. New Jersey, Hawaii, Maine and Massachusetts, all forbid physical punishment. Some cities, New York, Chicago,

St. Louis, Washington, D.C., Pittsburgh, Baltimore, Atlanta, San Francisco, Berkeley, Milwaukee, Boston, have outlawed physical punishment, while other cities in those same states permit it. The Los Angeles Board of Education outlawed it for three years, then renewed it in 1978. Parochial schools follow public-school laws of their states; some beat, others do not.

Millions of children spend school hours in an environment in which a teacher, a principal, an athletic coach, a playground or lunchroom supervisor has the right (with parental permission in some states and without permission elsewhere) to whack with a wooden paddle, hit with a ruler or spank with a rattan switch the hands, palms, face, legs, feet, buttocks, thighs, shoulders and arms of a student for misbehavior or unruliness. This is done under a doctrine called *in loco parentis*, which holds that "a teacher takes a parent's place and can spank a student if he does something for which his parent conceivably would whack him."

Laws that permit physical punishment call for the use of "reasonable force" and no "inappropriate instruments," but teachers have hit children with brass knuckles, leather belts, rubber hoses, baseball bats split in half, and have punched them with their fists, resulting in broken bones, broken eardrums, nosebleeds and bleeding welts. Teachers also painfully yank ears and pull hair. Rules require that beatings be done in private by one adult in the presence of a witnessing adult in a manner that "does not injure, degrade, or disgrace" a student, but many adults hit children on a playground, in an auditorium, gymnasium, classroom or lunchroom in front of their peers. A friend in his seventies still remembers with anger the teacher in the Chicago schools sixty years ago who sadistically gave a painful yank to his hair every time she walked past his desk. After taking it for many weeks, he was so goaded by pain that he stood up one day, clenched his fists at her and said, "Don't you ever do that again." She never did.

A child's disobedience may range from being late in sitting

down when class is called to order, to talking during class, chewing gum, not standing quietly in a line, talking back to the teacher, poking another student, not turning in homework on time or using a four-letter word. One high-school student was so severely beaten by an athletic coach that his jaw was fractured—for forgetting to bring gym clothes. Elsewhere, a music teacher swung a twelve-year-old student so hard against the wall because the student and three other boys were improvising their own sounds to a song the class was singing that the student's collarbone was fractured. The student was awarded $4,500.

At any one time there may be as many as 1,500 lawsuits against teachers and school systems brought by parents whose children have suffered excessive, unfair, or bizarre punishments. According to Adah Maurer, Ph.D., founder of End Violence Against the Next Generation, a national organization based in Berkeley, many more lawsuits are threatened or initiated by parents, but settled before trial. Nearly a hundred beatings a day in American schools wind up in actual or threatened lawsuits. Thousands of assaults are never reported; schools don't want them to show on official records and parents fear reprisals against their children.

It is an astonishing and horrendous fact, that an adult criminal offender may not legally be beaten in jail, but a child may legally be beaten in school. Other countries long ago outlawed beatings of schoolchildren—France in 1887, Holland in 1850, Finland in 1890, Poland in 1783. State laws rather than federal statute outlaw it in Germany today. (It had been abolished but was reinstated during the Nazi regime.) Those who oppose corporal punishment in the United States believe that it could be stopped overnight if a federal law were passed denying financial subsidies to schools that permit it. Their cause was done great harm when President Jimmy Carter, touring Japan in 1979, laughingly told a schoolboy that he was often "severely punished with a paddle." One of the reasons he became President, Carter said, was because his teachers had been very strict and "inspired me to study."

In the United States we have a tradition of beatings dating back to Puritan days. A Texas judge, upholding the right of schools to paddle students, has said that "Nature has provided a part of the anatomy for chastisement, and tradition holds that such chastisement should there be applied." In April 1977, the United States Supreme Court ruled that there were safeguards in state laws to protect schoolchildren. Said Justice Lewis Powell: "The schoolchild has little need for the protection of the Eighth Amendment" (against cruel and unusual punishment).

The horror is compounded when we learn that most beatings are of the smallest and meekest children. High-school personnel are afraid to tangle with students who have grown large and can hit back. Los Angeles teachers have a contractual right to spank in elementary and junior-high schools but not in senior-high schools. In Idabel, Oklahoma, a thirteen-year-old student, enraged by a paddling, told the principal he would kill him. Grabbing the paddle with which he had just been beaten, he hit the principal's head so hard that the paddle broke. The principal was rushed to the hospital for emergency treatment and stitches. The student was turned over to the police.

Corporal punishment also reflects many of the racial and class problems of our society. According to the National Education Association, it is used more frequently against poor and minority children than against white, middle-class children. In 1970–71 Dallas schools reported 5,358 spankings. The following year when integration began, Dallas schools reported 24,305 incidents, a fourfold increase. The year 1978, in which Los Angeles teachers asked that corporal punishment be restored, was the year in which integration started via school busing.

"What's so tragic," says Lillie Kogan, a retired Los Angeles teacher, "is that kids who are beaten are frequently those who especially need caring contact at school because they are not getting any at home." Beatings in schools even encourage beatings at home. Many parents have the attitude that "if teachers and principals do it, why can't we?"

Children need the blessing, not the bestiality, of touch. Nearly every touch of hard weapon against soft flesh not only welts the skin but also lacerates the psyche. Every one of us pays a price for school beatings: the child who suffers damaged self-esteem and self-respect and is motivated to get revenge; the beater—a Task Force of the National Education Association says "repeatedly and righteously inflicting physical pain is likely to be detrimental to a teacher's mental health"; the public, both in the spending of our tax money to settle claims against schools and in the diminished quality of our lives when we live in fear of young people growing up angry at the world.

The problem of discipline in schools is certainly not one-sided. There are more than 100,000 assaults on teachers each year. I feel great empathy for school personnel who go to work every day with their stomachs in knots from the fear of being attacked in classrooms, corridors or lounges. Columnist Ellen Goodman, in an article "America's War Against Its Children Makes Monsters of Us All," points out that "Teachers are battered almost always by older children." Laws that allow teachers to commit violence on small children give implied permission to older children too to be violent. Also, older students who are physically large may appoint themselves the protectors of smaller children who cannot hit back. Such laws have never solved disciplinary problems and they never will. Temple University has set up a National Center for the study of Corporal Punishment and Alternatives in the Schools. Its director, Irwin Hyman, Ph.D., says, "Americans are more willing in many cases to touch punitively than to touch lovingly. My research shows that extremely punitive (verbal and) physical contact results in lower achievement and creates greater discipline problems."

As long as teachers have the option to punish by painful touching, not enough emphasis is being placed on exploring alternative forms of discipline that can be healthy, constructive and ego-enhancing. We could be experimenting with many alternatives—peer review boards in which students accused of misbehavior come before peers (peer groups sometimes impose harsher penalties than adults do); deprivation of privileges such as attending athletic events, dances and school outings; a PACT Committee (Parents, Administrators, Children, Teachers), which sets up a system of reconciliation, perhaps an ombudsperson, based on the Swedish system, to whom teachers and students each tell their respective sides of an incident.

Adah Maurer suggests that, at the moment when, say, two pupils are in physical battle, a teacher separate them and then if necessary firmly hold the arms at the sides of one combatant until the heavy breathing subsides. After that the teacher can give instructions about where to go and what to do until their differences can be handled calmly or referred to the proper place for adjudication.

New approaches to discipline would redefine what punishable behavior is. What seems like misbehavior may sometimes be a student's creativity coming out in untraditional ways. To return to the example mentioned earlier in which a student's collarbone was broken when he and classmates added their own vocal flourishes to a song the class was singing, if creativity had been prized in that situation, the teacher might have said, "Let's all hear those improvisations." He might then have gone around the room encouraging all the students to create their own improvisations. That experience could have been of greater value to students' musical education than continuing with the song.

A similar incident occurred in a nursery school where I worked. The teacher, recognizing the creativity of children who spontaneously veered away from the classroom song, encouraged their impromptu compositions, which led to hilarious numbers such as "I like tuna fish for lunch" and "You smell nice, teacher." That incident led to occasional miniature operettas in which the children sang out their feelings.

As we explore new ways to discipline, Sidney B. Simon, professor in the Center for Humanistic Education at the University of Massachusetts, believes that we should give priority to touching, because it is such a powerful tool. He says, "If touching were permitted—even tolerated—in our schools, how much less grief and anxiety and deep feelings of inadequacy we would find in our young people." He is convinced that students who repeatedly create discipline problems by shoving, pushing and tripping others do so because they are embarrassed or self-conscious about wanting to make touch contact and so they do it in a malicious, teasing or combative manner.

Here are some "skin strategies" Simon uses in his classes.

> *Massage Train.* The group forms a large circle around the room. People turn to the right and put their hands on the shoulders of the person there. Someone has his or her hands on your shoulders and you have your hands on someone else's shoulders. Bring some comfort to the tired shoulders of the person you are touching. Make your hands healing, comforting, soothing, full of caring.
>
> There will be some verbal discharge of anxiety but insist that they let their hands do their talking. The students will settle down. Some of the more embarrassed may show momentary stress by massaging too hard or tickling. Repeat the instruction that hands must be healing and comforting. I usually inject a funny idea by saying: "This is the way we're going to begin all school board meetings from now on."

I don't let the exercise run so long that people get tired reaching out. Then I say, "Slowly, let your hands come to a rest but don't take them away yet. Cradle that person's shoulders in your caring hands. And then, slowly, take your hands away."

The next direction is, "Now turn to the person who gave to you and give to that person." They get right to it, with less embarrassment than on the first round. They will be telling you, non-verbally, that this is something they have been wanting and needing for a long time.

At this point you probably can go right into teaching your first subject for the day. The total exercise has taken a few minutes. Skin-hunger caring will have been brought home to them. You can see a noticeable tenderness in the group towards each other.

Temple Caring. This strategy comes after you've done a few days of massage trains. Divide the class in some way; my preference is to have people quickly pick partners and then measure heights. They form two groups, shorties and longies. However, anyone who has always been a shortie can ask to be a longie. Then say, "Shorties, sit in a chair. Keep your eyes closed until I tell you to open them. Longies, stand behind a shortie and touch him or her gently on the top of the head." I add an important concept when I say, "Help each other find the people who aren't covered yet." I teach my students to snap their fingers if they are standing next to a shortie who hasn't been found by any partner. The "cricket" noises guide people who are still looking for a partner. I've done this in a gym with 500 people; it's incredible how quickly people rally to help each other find a partner. Live crickets never sounded so sweet.

If there is an even number of students, I don't participate. If there is an odd number I always take a partner, which is my preference. I don't like taking students through something I'm not willing to do myself.

I then say, "Now, longie, stand behind your shortie. Shortie, keep your eyes closed. Later on you'll find out who your secret partner is. Longie, lean your shortie's head gently against your body and begin to give him or her one of the most gentle, caring, aware temple strokings you will have ever given anyone." And they do. You will see some of the roughest kids caring tenderly for others.

One of the advantages of the shortie and longie pairing is that it is usually men who are longies and women who are shorties. Caring for people of the same sex is still so taboo in our culture that you would be wise to avoid it in the beginning. But inevitably there will be some pairing, as in the massage train, in which a guy gives to a guy and a woman gives to a woman. I usually defuse the situation with something direct and yet light. I say, "Now, listen, it's okay if you are a guy with a guy. No one is going to accuse you of being a homosexual. Later on you might even learn about the sexless nature of this kind of comforting and caring."

Give about three minutes to the temple rubbing and then say, "Let your hands slowly come to a rest and cradle that head, ever so lightly between your hands. Let your partner feel safe there with the warmth and tenderness and healing quality of your good hands." Then tell them, "Slowly take your hands away and come around the front and let your shorties see who was there for them." There is a roar of delight and surprise when the shorties discover who was comforting them. Then I quickly say, "Okay, shorties, you get a second one." Whereupon every shortie cheers. They are told to close their eyes again and settle down. The longies are instructed to go off and find a new partner and repeat the sequence. Finally, the longies get to sit in a chair and receive two temple rubs from two different shorties.

When this is over, the class would really profit from a chance to share what they experienced. You will be delighted with the kind of talking they do. You will get tremendous satisfaction from the awareness of kids you'd least expect to have it. You will also see the calming, soothing effect that skin-hunger caring has for students. The class work will go more easily, and you will have fewer discipline problems, even in that single class period.

However, because of our culture and its ground rule that any touching is sex, you would have more difficulty if you were to repeat these exercises the next day. People will have had overnight to get back to feeling uptight and they will have had 24 hours to get their guilt feelings back in force. "All that pleasure must be immoral," some of them will think. Consequently, you'll have to quote the studies about the orphanages. [Simon is referring to the René Spitz study described on page 26.] You'll have to lead a discussion about the kinds of problems which result in our society from its use of only two modes: no touch at all or let's go to bed, baby.

Head and Body Tapping. Demonstrate this strategy with a partner. Stand behind your partner, curve your fingers, and tap gently on the person's head, like raindrops falling. Check with the person receiving the raindrops to learn if they are too hard or too soft; no one should suffer any discomfort. When everyone has a partner, encourage the whole room to get into the same rhythm. Everyone is tapping gently to the same beat. Then suddenly you say, "Take your hands away, at once." Everyone will laugh—the receivers because their scalps tingle and the givers because their fingertips feel so alive.

Then you say, "Okay, once again. Tap, tap, tap. Now faster and faster and zoom—take your hands away. Make your hands flat and let's start gentle body slapping on the shoulders. Slap, slap, slap. Increase the speed and a little of the strength but don't hurt your partner in any way. Now take your hands away."

Ask the receivers how they feel. You will find they feel more alive and energetic.

Next I tell them: Flatten your hands and start on your partner's shoulders; go down the arms, over the hands, and back up again. Slap, slap, slap more and more rapidly until you reach the shoulders, then take your hands away—zoom.

For the last body tap, start on the shoulders and do the whole back, never hurting but stimulating the blood and the energy flow. Take your hands away suddenly. Now partners, change places.

Back Massages. Here you need a room where the chairs are movable, but nothing more. I use a typical university room. We pile the chairs up or push them to the side. People get their jackets or sweaters and lie face down on the floor. I encourage them to relax and give their weight to the floor. I get them to think about their day—what it's been like. Those who seem to be having a rougher day usually get to be the first takers. Those who are experiencing a gentler day give first. Eventually, though, everyone takes and gives. Reciprocity is important.

Back rubs are almost always given anonymously, although students soon learn to recognize hands and styles of massage. I feel the anonymity demonstrates that 30 different people can be there to meet skin-hunger needs, not only the ones who are cute or prestigious. Dozens of students who were full of self-put-downs on the first day of class now know that they are important and special to almost everyone else in the class.

Lots more could be written about back massages but suffice it to say that it is one of the most important strategies a school system can employ. And it's free. I can't think of another effort which does more to build a feeling of community and belonging than for students to be able to ask for a back massage when they are feeling down or tired.

Summing up, Simon says to teachers:

Think about what happens in your school cafeteria when a kid drops a tray and the dishes break and the food goes sloshing across the floor. In most schools, the students shout and hoot and whistle and stomp their feet and clap

> their hands and pound on the tables. They do everything they can to add to the humiliation of a student who is already embarrassed. When I hear that hooting and hollering, I know I am in a school where there is rampant skin hunger.
>
> In a school which meets students' skin-hunger needs and where comfort and caring go on throughout the school day this is what would happen if someone dropped a tray. Two kids would get up out of their seats and help clean up the mess. A third would go back in line and get another tray of food. A fourth would stand behind the student who dropped the tray and give a gentle back rub to ease his tensions.
>
> I have never seen anything which has more potential for building community and preventing discipline problems than skin-hunger work. It is almost magical.[1]

I interrupted the typing of this to attend my jazz aerobic class at the Hollywood YMCA. It was a coincidence that the teacher had us pair off and do similar "skin strategies." My partner, a woman new to the class, was uneasy about participating. I urged her to do so. Afterward she exclaimed, "Wow! I didn't know what I've been missing!"

It would be easy for schools, beginning with the earliest grades, to incorporate the idea of touch into classroom learning. In "Show and Tell," now a common technique in lower grades, a child brings an object from home and tells about an experience with it. "Reveal How You Feel" could have a child telling how he felt when he was comforted by a friend or parent or teacher. Humorist Sam Levenson, who was once a teacher, says, "It is more important for the child's first reader to say 'Love, Dick, love' than "Jump, Dick, jump'. It is just as easy for a child to learn the word 'kiss' as 'miss,' 'hug' as 'bug.' "[2]

Teachers of English could assign essays in which students tell what messages they receive when a friend touches a hand or puts an arm around their shoulders: "I like you. You are a worthwhile person. I am your friend. You can count on me when you need me."

With young people from so many varied ethnic backgrounds entering our schools, educators can use discussions of touch in students' ancestral lands as lessons in geography, sociology, value clarification, cross-cultural behavior—teaching that democracy respects diversity. Even within the American culture we have subcultures with different attitudes. Mike Fire and Constance Baker, Oklahoma teachers of American Indian students, write in the *Journal of Nursing Education* that

> touching as a positive reinforcement is not as acceptable to the American Indian as to the general student body. Touching is usually reserved for a tribal member who occupies a high position in tribal hierarchy. Touching generally occurs between members of the same sex and when there is an unwritten agreement that allows for contact, in happiness, sadness, sharing. There is no contact at all, except for handshaking, within a family. The Indian believes that to touch someone violates his body boundary and takes something away from him as a person.

The teachers also experienced that "prolonged eye contact with another person," which can be considered a form of intimate contact, is "thought of in the Indian culture as a disrespectful act. An Indian student's characteristically downcast eyes as a sign of respect can be difficult for a teacher to accept."

At Brainerd Community College, in Minnesota, Joseph Plut, Jr., an English teacher, is known as the Mad Hugger, a custom he started after he heard Leo Buscaglia lecture on love. "It's not just physical, it's more spiritual," he says. "It's like hugging another person's spirit. It means 'I'm accepting you and you're accepting me.' " Looked upon at first with some embarrassment and ridicule, campus hugging has now caught on with many students and faculty doing it. To show how hugging is infiltrating, a friend sent me this description of a ceremony of the Los Angeles Community College District:

"At the recent investiture and presentation of the Presidential Medallion, Dr. Leslie Koltai (Chancellor) put the medallion on a ribbon around the neck of President Thomas L. Stevens, Jr. Still with the medallion in his hands, Dr. Koltai paused, quickly stepped forward and impulsively threw his arms around President Stevens, giving him a big bear hug and kissing him on the cheek. A sort of shocked hush fell over the audience and then—as one—they applauded joyously. It was, indeed, a rare moment."

As tactile contact between teacher and student becomes more acceptable, might there be abuses, especially against young students? Some education experts stress concern that encouraging touching in schools, "without regard to the power relationship between the touchees," can be dangerous. Adah Maurer, mentioned earlier, warns that boys and girls eight, nine or ten years old may find embraces offensive, even babyish. Children may "wriggle their way out of a grandmotherly embrace and wipe away unwanted kisses with exaggerated disgust."

A male teacher in an elementary school is concerned about seeming seductive when he touches his pupils. He finds "a handshake is a desirable tactile contact, even with small children, who may giggle with delight at being accorded such grown-up behavior. A handshake shows warmth and respect for the child's accomplishments and yet does not coerce the child" who is inevitably in the less dominant position in the power hierarchy.

Edgar Dale, educational philosopher, urges us to stop "adjusting students to a world that is dying and to prepare them for a world that is being born." In the world now being born, more teaching will be done with machines. Pupils will be sitting in front of consoles, each working at his or her own pace.

A machine can teach, correct, reward with a word, but no machine can give a student that delicious spark of approval that a teacher's pat or handshake can give. There is no greater motivator for learning than to know that one is cared for and respected. Says Sidney Simon: "It is amazing how much work we get done in school when we care for skin-hunger needs first."

12 Touching in Business and Politics

This book began as a magazine article, which was reprinted in *The Los Angeles Times*, Sunday, November 27, 1977. The article opened with a report on a study at Purdue University (mentioned in Chapter 7, "Touching Strangers") in which librarians touched students who were returning books. The students were not aware that they had been touched, but later reported to researchers that they were having good feelings about themselves, the library and the clerks.

The owner of a chain of restaurants read *The Times* piece and decided, "I'm going to do that in my business." A year later I met him at a party. When he heard that I was the author of that article, he exclaimed joyously, "You don't know it, but you're helping my business to grow!" He told me that he instructs employees to lightly touch customers whenever it's appropriate—giving change, handing out or retrieving menus. While he has not conducted a formal study, he feels certain that the good feelings that customers have about his restaurants, as shown by repeat patronage and a growth in the chain, are partly the result of these pleasant touches.

Increasingly, business people are using the sense of touch to promote sales and to enhance customer relations. During a recent Christmas season, the advertising theme of the May Company Department Stores in Los Angeles was "Touch." A series of ads described The Christmas Touch, The Silken Touch, The Touch of Class, culminating in an ad on Christmas Day wishing customers "the warmth and love of Christmas." Illustrating the ad was a detail from Michelangelo's ceiling in the Sistine Chapel showing the outstretched finger of God giving life to man. The ad read: TOUCH The Greatest Gift of All.

Bell Telephone Systems has a television advertising campaign based on the catchy musical jingle, "Reach out and touch someone far away." That musical line keeps running round in my head—exactly the way Bell hoped it would! A Bell executive told me that these ads, "the most successful ever," have "great emotional appeal." Many people call to ask for the words to the song.

Esoterica Dry Sky Lotion, in a commercial, has a model saying, "I want to be touched, not just looked at." A model for Natural Wonder Cosmetics says, "Being close to people is important to me. So my makeup has to look . . . and stay . . . fresh and touchable."

An advertising slogan for the Santa Monica Bank asks, "Have you hugged your money today?" A research firm, Marex Marketing, experimenting with a plan to have clerks hug each person who buys a Plantable Animals hanging—plant shaped like teddy bears, swans, and owls—stationed young men at street corners in a dozen cities to determine how the public would respond to hugs from strangers. More than half the women who stopped were willing to be embraced when the male researcher asked, "Do you mind if I hug you?" ("Did you say 'hug' or 'mug'?" asked one cautious woman.)

An enchanting commercial for the San Diego marine playland asks, "When was the last time you were touched by the wonder of a cuddle—a hug—a tickle—a squeeze—a back rub—a nuzzle—a caress—a kiss . . . all in one day? Sea World. A touching experience." Interspersed among the words were scenes of men, women and children nuzzling a whale, a dolphin, a starfish, a deer, ducks, and one another.

Mail-order companies sell us tactile comfort with such items as "Worry Stones," smooth rocks over which we can run our thumbs and fingers to calm us when we are tense. Business journals advertise The Executive Sandbox, a decorative box in wood finishes to match desk decor, containing sand that an executive dribbles through his fingers as he talks on the telephone, dictates to a secretary, or sits ruminating.

I once thought of such items as nonsensical fads but I have changed my mind. I have tried them. The rhythmic touching does have a hypnotic, calming effect. The principle behind such tactile products is much the same as that used in autogenic training, where one is taught to repeat rhythmically a physical action in order to discharge bodily tensions.

Some companies are even responding to the tactile problems of a special segment of the population—left-handed people. About thirty million people, among them Gerald Ford, George Bush, Caroline Kennedy, Robert Redford, Cary Grant, Ringo Starr, Richard Pryor, Bob Dylan, Jimmy Connors, Martina Navratilova, Reggie Jackson, Greta Garbo, Erma Bombeck's husband, are lefties. So were Leonardo da Vinci, Michelangelo, Benjamin Franklin, Nelson Rockefeller, Napoleon, Mark Twain, Pablo Picasso, Harry Truman, Cole Porter, Charlie Chapman, Marilyn Monroe and Judy Garland. To be left-handed in a world of right-handed products is to constantly experience small touch discomforts and irritations of which right-handers are unaware.

Observe a left-handed child writing in a spiral notebook designed for the right-handed, how her hand doesn't sit comfortably because it is hitting the metal spiral. Watch a left-handed child using a wall pencil sharpener made for right-handers, or trying to open cans with openers that have handles on the right side, or dishing out ice cream with a scoop on which the releasing trigger is designed for right thumbs. Television knobs are usually on the right side; so are controls and tone arms on stereos. Cords on telephone receivers are connected to the left side, freeing right hands for writing. The thread on corkscrews is spiraled for right-hand turners; lefties need the thread turning the other way. A "leftie" humorously accuses the world of another "ism" of prejudice. He says we are guilty of "sideism"—discriminating against people's left sides.

Business is beginning to change all that. An organization, Lefthanders International, has drawn up a Bill of Lefts which it sends to manufacturers suggesting left-handed designs. "When we started," says Jancy Campbell, executive director, "there were four mail-order companies selling left-handed products; now there are more than twenty-five." There are also twenty-five LeftHanded Shops across the country. When I talked with Ms. Campbell, I gave her an idea for one more left-handed scissor to be added to the barber, kitchen, sewing and fingernail scissors already available—garden shears. (Lefthanders International, 3601 S.W. 29th Street, Topeka, Kansas 66614.)

In one sector of business, touch is a major part of the "product" we buy. Barbers, beauticians, hairdressers, masseurs, manicurists and pedicurists are in "hand-to-body" businesses, which are selling touch along with their skills. The size of their clientele and the amount of money they make often depends upon how customers feel about the tactile contact between them. A friend lapses rhapsodic about the "golden hands" of her hairdresser, who is almost as important in her life as the husband who earns the money with which she pays the coiffure artist with the delicate touch. Another woman drives forty miles a week, round-trip, for the services of a particular manicurist. I asked her what was so special about that person. She said: "Oh, her lightness of touch when she cuts my cuticles, the way she pulls and circles each finger, the creamy massage from my hands to my elbows, that final loving pat on the top of my hand when she finishes."

We treasure these people, because they are surrogate "parents," who nurture us as if we were babies. No matter how famous, important or rich we may become, there remains in each of us the child we once were; we like it when someone's hands fuss over our physical beings, even if we are paying for it. How blissful it can feel to lie back on a chair with your head resting on the shampoo basin and have warm fingers moving about your hair and scalp. It is with a delicious feeling of freedom from care, of "coming home," that I become, at my weekly visits to the hairdresser, a "child" being taken care of by a "mommy."

For too many of us, a visit to beauty parlor or barber shop is the only occasion for being touched by another. After reading my article on touch, a hairdresser in a small town in Idaho wrote to me:

> Sometimes when I arrive in the morning, a few elderly women are lined up outside my door looking as if they'd just come *from* a beauty parlor. Some come twice a week for shampoos and sets they don't need. I know they are there more for physical and human contact than for the hairdo. I feel sympathetic; I let the shampoo go on as I massage their scalps tenderly. I touch them frequently around the shoulders. When I finish combing them out, I give them a light squeeze on the back of the neck.

Poor people suffer a subtle kind of deprivation. They need and would enjoy touch as much as the rest of us do, but they can't afford to "buy" touches as the more affluent can. I have an idealistic dream of two things I would do if I had lots of money—one is to provide free hairdos and manicures for impoverished men and women; the other is to organize Mommy

Vacations, a program that would provide weekends for poor mothers who never get a day or night's relief from child care, where someone would serve meals to them and massage, shampoo, manicure and pedicure them.

Most of us—as customers—welcome business touches because they evoke pleasurable sensations. But, as employees, some of us find touch to be the cause of psychic pain, fear and anger. Such touches are defined as "sexual harassment." As more women enter the world of paid work, more are reporting unwanted touches, grabs, pinches, pats and squeezes from male bosses, clients and colleagues. A woman's status makes no difference. White-collar executives and women who own their own businesses, as well as pink- and blue-collar women in factories and in service jobs, are all being pawed and pressed against. *Where* and *how* they are pawed and pressed may depend upon the man's position in the company hierarchy. I sat in once on a meeting of the National Organization for Women at which women talked about such experiences. They gave these names to the types of advances and the places where they are made.

A clerk or maintenance man may operate in "grope alley"—the narrow space between rows of cabinets in the file room. "That's where the guy gropes for a feel when you're stuck in the tight space between an open file drawer and the cabinet behind you." Women who eat in the company's cafeteria may get snared by "lunchroom lust." "That's the knee-to-knee or thigh-to-thigh maneuver under the table, while above the table, innocent eyes look at you and pretend they can't imagine who's doing that to you."

Middle-management men who may have cubicles but no private offices make their attempts at physical contact under the guise of "water cooler camaraderie." "As you're standing at the water cooler," one woman said, "filling a paper cup, a hip gets brushed against your hip or a shoulder presses against your breast. The guy laughs at how cleverly he maneuvers this so it seems 'accidental.' "

Those high up enough to rate pitchers don't need to use the cooler; their method of attack is "carafe contact." "As you pour water into a glass, your boss closes his hand intimately over yours and 'helps' you to pour," explained an executive secretary.

Women employed on a production line contemptuously give out a secret annual *3-F Award*—for the "Factory's Foremost Fucker."

Male superiors have probably victimized and exploited working women with unwelcome advances ever since the first man hired the first woman. Labor history is thick with incidents reported by helpless women who were the first to work for pay at the start of the Industrial Revolution. Many lived in fear of losing their desperately needed jobs if they resisted the grab at the breast, the reach under the skirt for a squeeze of the thigh or a clutch of the pubic area, the pinch of the buttocks, even acts of rape, by bosses, foremen, supervisors and job agents. Years before she wrote the classic *Little Women,* Louisa May Alcott wrote an article entitled "How I Went Out to Service," telling how she was molested by an employer in whose home she worked, a Reverend Josephus. She quit, later writing that her heart had "suffered many of the trials that wound deeply yet cannot be told."[1]

Women could be fired because their feminine presence was considered a physical temptation. In 1854 three young women were hired as copyists in the U.S. Patent Office. A supervisor threatened to fire them because they were females. The Commissioner of Patents wrote to Robert McClelland, Secretary of the Interior, trying to save their jobs. McClelland answered, "There is such obvious impropriety in the mixing of the sexes within the walls of public office that I have determined to arrest the practice." One of the discharged women was Clara Barton, who later founded the American Red

Cross. (In an unusual example of role-reversal the Bell Telephone System, until a decade ago, never employed male operators. Its long-standing rule was that "switchboard positions are necessarily very close. This fact, plus the need for reaching to establish connections on the adjacent positions, makes physical contact between operators absolutely unavoidable." Now they have male operators, but telephone equipment is so modernized that there is no physical contact between operators.)

Working women still suffer these same indignities but today, unlike Louisa May Alcott's day, the "trials that wound deeply" can be told—to the Equal Employment Opportunities Commission and to various Human Rights commissions, through which women now have legal means to seek redress for tactile and verbal sexual pressures on the grounds that such behavior deprives them of their civil rights. Resisting overtures, women are in danger of being fired, often on false grounds of incompetence or inefficiency, of being deprived of promotions and raises, of suffering loss of job benefits and overtime pay. Thousands of women say they work fearfully under the threat of such penalties if they don't "come across." Should they quit, they have been unable in most states to collect unemployment, a situation that women's groups are seeking to change. The commissions have backlogs of cases awaiting hearings, some of them filed after women have been fired, others while women are working uncomfortably every day on jobs with punitive male superiors against whom they have brought suit. Lawsuits have resulted both in cash awards and reinstatement into jobs for some women. "Yeah, I won," says a victor, "but I'm made to feel like a pariah on the job, so I'm looking for work elsewhere."

Few men have lost jobs for doing the harassing; many women have been fired or quit because they were harassed. Jane Fonda told an audience of office workers, "As a secretary, I was once fired for not sleeping with the boss." A male employee of a food company "put his arms around one of the women employees, lifted her off the floor, and moved his body back and forth against her" in a simulated sex act. After a struggle the woman broke free. A supervisor saw the incident and asked a union steward to take action. Management was willing to dismiss the man, against whom there had been complaints before. A union committee called the incident "sexual horseplay" and rescinded the discharge (but they suspended the man for a year). Robert E. Quinn, Ph.D., assistant professor of public administration at the State University of New York at Albany, a noted researcher into office affairs, says, "The female is twice as likely to be terminated as is the male."[2]

For a woman, complaining to higher-ups, who are nearly all males, rarely does any good. Their response frequently is, "Oh, he's just kidding around. You take it too seriously." Or, "Aren't you mature enough to handle it?" Higher-ups themselves may have "executive suite-hearts," a description used by *Dun's*, a business publication. Apparently an "old boy's club" permeates business; men are protective of one another, reluctant to call other men down for something they themselves may be doing. Linda Singer, a Washington, D.C., attorney handling harassment cases, believes that judges are reluctant to "recognize these as sex discrimination, because they fear it could become a cause of action to blackmail men."

A man may think of touching a woman at work as nothing more than an amusement, a tactile "kibitz," a question to see how far he can go and whether she will respond, a test of his charm and desirability, a hostile way to humiliate and unnerve her because "women don't belong here anyway," or the acting-out of anger he feels toward his wife or lover. A woman feels his touch as if it were a slap, a negation of all she has worked hard to achieve, the long hours of schooling and, if she is a homemaker and mother, the many adjustments she has had to make in her life in order to be on the job at all. If the unwelcome intrusion of a man's touch can assault her self-esteem and bodily privacy she may well feel despair, thinking: "Is this all I'm worth, a sex object? What about my brain, my competence, my training to be in this job, my dedication to what I'm doing?" Says one woman who undoubtedly speaks for many, "Can you imagine what it is like to love your work, the surroundings, the company, and yet walk around with a nervous flutter in your gut because of one insensitive man who's in a position to take all that away from you? I distort myself trying to avoid running into him in the hallways, on the elevators, on the stairways."

The unwanted "career caress" lays heavily on women. It carries an implied obligation to reassure men of their virility and desirability. A woman is supposed to provide a man with self-esteem by how she responds to him. This is one more burden on the male-female relationship, similar to the one in which women are *supposed* to have orgasms, not for their own pleasure or because their bodies need orgasm, but to reassure the man of his prowess as an orgasm-producing lover.

At the instant a male superior or colleague touches a woman in a seductive manner, she may be so unprepared that she doesn't know what to say. Women berate themselves afterward for not handling the situation better. Dr. Mary Walshok, a sociologist who studies the career patterns of educated women, believes that this "problem comes up for almost every working woman" at least once and that women ought to be prepared with what they will say when it does happen.

Asked by a reporter how they would handle a Lothario on the job, women made these suggestions, some in jest, some seriously:

Pretend that you do not understand what he means. If a man has to keep asking you, pretty soon he gets tired of asking. He's usually not that sure of himself anyway, and after a few times he'll stop trying.

Treat the whole thing as a joke. If you pretend you think he was kidding, it helps save face for him, too.

Be shocked, outraged and hurt all at the same time. Then start to cry. But be sure to do this in private. You don't want to create a scene in front of other employees. He'll probably dislike your crying as much as he dislikes his wife's crying and he won't approach you again.

Tell him how flattered you are—remember he's your boss and you don't want to hurt his male ego—but why should we spoil such a good working relationship? The business journal *MBA* suggests somewhat the same thing. It advises executives to consider whether they might spoil a good working relationship by getting romantically involved.

Tell him you have too much respect for his wife and family to hurt them. If he is unmarried, tell him you have a boy friend or a husband. A college instructor said, "I usually say, 'I like you but I don't think my husband would approve.' Let him down easily. It's no good to clutter your working life with enemies."

Try the shock approach. Tell him that you are in the middle of a passionate affair and that that is all you can handle. You can even strengthen it by saying you are in the middle of a passionate love affair with some other man in the company. He could think it might be *his* boss.

Memorize a few one-liners such as "My psychiatrist says I'm not ready to get involved in another love affair right now, but I'll let you know how I feel about your suggestion when I come out of analysis."

Michael Korda in his book *Success!* advises career women:

> It is not a bad idea to keep a man's picture on your desk, provided he looks suitable, solid and of the right age. One young woman put a photograph of her sister's husband, a football player, on her desk. Another, more daring young woman put the picture of the chairman of the board on her

desk, in a silver frame. Since the company she worked for was one of those huge financial institutions with thousands of employees, none of her immediate colleagues or superiors felt in a position to ask whether she was in fact related to the chairman.

These suggestions require game-playing maneuvers offensive to feminists who believe that women should come on straight when expressing their resentment over sexual advances. Groups such as Alliance Against Sexual Coercion (AASC) in Boston, the National Organization for Women, and Working Women United Institute (593 Park Avenue, New York, N.Y. 10021) urge women to bring harassment into the open and aid them emotionally and financially should they bring suit.

I have mixed feelings about rushing into legalities before other means are explored. I teeter between worrying about the trauma of lawsuits and the stigma they can put on a woman's work record in applying for future jobs, and feeling empathy for women whose breasts, buttocks, abdomens and pubic areas are touched by an insensitive man when they are alone in an office or other workplace. When I was thirteen years old and pretending to be sixteen to obtain Saturday jobs, I worked in a dress shop and spent part of each day alone in a back room folding cardboard dress boxes. Once, the elderly owner came in and put his hands on my breasts. I was terrified, torn between needing the one dollar that I was paid for the day and wanting to run out. I retreated to another part of the small room. He did not attempt any further touch.

But I can identify with the crawling sensation women feel when someone on whom they are dependent for their jobs does this to them. It is a dreadful invasion of one's body, a psychic rape. Today if this were to happen to me I'd say forcefully, "Thank you for the compliment. I prefer that you value me for my work, not my body, and that you not touch me again." I would do it the *very first time* a man touched me in an offensive manner. If we do not protest a disturbing act the first time but plan to protest it some future time, we subliminally give permission to the other person to try it again. I call this "creative confrontation." Some people have a fear of confronting, believing it must necessarily be angry. Confrontation can be quiet, friendly and not angry, and still be effective.

I admire the way in which an older woman newly returned to the world of work confronted her boss after considering legal action and then deciding she would try to resolve the situation herself in a friendly manner, thus keeping her job and his respect and allowing him to save face. She did what Lawrence D. Schwimmer, a business consultant specializing in interpersonal relationships, advises; she wrote a note:

> I like my job and know you like yours. We share a common goal in wanting the company and our department to do well. However, I find myself working with anxiety because of some incidents between us.
>
> I have worked hard to educate myself to be worthy of my job. I need to earn the money because I want my children to have what they need—just as you do for your children.
>
> My goal is to do well on my job, to be happy doing it, to advance in the company, and to achieve a happy family life. It is hard to do this if I have to be working with trepidation that my progress depends on accepting overtures from anyone. I want to be valued for my skills and intelligence and competence.
>
> I would like it if we could work together in harmony and mutual respect as colleagues. Can we be friends?

Her boss did not answer her directly, but he showed by his demeanor and tone of voice that he respected her. They worked together for a couple of years when he helped her earn a promotion by writing a strong letter of recommendation.

She says now, "The option of legalities was always open to me. I'm so glad I tried this way first."

An important aspect of business romance is that jobs provide a legitimate ground for finding relationships. Many romantic contacts are made through one's work. Who among us has not felt the rush of blood, the sudden inflamed rise in temperature, the rapid heartbeat, the sweaty palms, the catch of breath, the delicious heady sensation of being about to take off and float across the room, when our body or hands have accidentally touched someone while we were looking over papers, handing someone something, or unavoidably squeezing against someone in a crowded elevator? We worked better because of it. So there often are desirable and respectful sexual overtures men and women make to those who appeal to them and with whom they might develop brief or ongoing relationships.

Job-related romances will increase, especially as the number of single people grows. Until recently, corporations preferred married executives, believing that marriage and family were stabilizing influences. But companies are finding that single or divorced executives have more time to work evenings and weekends; they are freer to travel and to move to other cities. All other qualifications being equal, some corporations are beginning to favor single men over the married man whose neglected wife complains, "Do you have to spend so much time at work?" Psychotherapist Maryanne Vandervelde, a corporate wife herself, says corporations have put their seal of approval on divorce and singlehood. As more single people are thrown into job situations with no wife or husband to impede after-hours liaisons, and as more male and female executives, married or not, travel together to out-of-town conferences, overtures for sex by touch or verbal invitation or both will increase.

If both parties are agreeable, why should businesses object to company romances? Because they may lead to jealousy, gossip, reprisals and sabotage that affect productivity and efficiency. Robert Quinn, in the February 1978 issue of *Human Behavior Magazine*, says that there are three basic types of office romances: the *fling* (usually a brief relationship to satisfy a physical yen); *true love* (in which a man and a woman care about each other and have a long-term relationship similar to marriage); the *utilitarian arrangement* (in which a usually superior male "derives ego benefits from courting an attractive female who is in it for the job rewards").

No matter how much romancers try to keep their relationship secret, they tip it off by these behaviors:

Longer-than-usual conferences behind the closed door of one or the other's office, usually the male's.

Increased business trips together out of town.

Occasional mutual disappearances for a whole afternoon or taking prolonged lunch hours.

Scheduling summer vacations at the same time.

Standing and chatting a longer time, supposedly about company matters, than is normally done.

Working late together when no one else is doing so.

Prolonged goodbyes at one or the other's car in the company parking lot after work.

Romancers who are on lower career levels and don't have private offices but do have high libidos give themselves away by being caught "embracing in closets, kissing in supply rooms, or fondling in the parking lot."

Other employees may not be bothered by a company romance that supplies titillating gossip or speculation—until one or the other lover neglects work duties that are then loaded on to others. Resentment rises if a lover acts as a "spy" and reports back to the other, who may be a superior; if the lovers are habitually late getting to work or coming back from lunch; or if one, the woman usually, gets privileges, raises, promotions no one else gets, as a result of her romance.

True love brings out the best in all of us. Robert Quinn says, "When two participants are perceived as being sincerely in love, the relationship has a different meaning than when the male is perceived as always looking for the next conquest and the female is thought of as being on her way up in the organization."

Among the reasons business gives for resisting the hiring of women is that their presence can be a sexual temptation and interfere with productivity. Shades of Clara Barton, nearly 150 years later! The fear of sexuality is an excuse. "The ultimate purpose of touching restrictions," according to psychiatrist Roderic Gorney, "is to maintain the order of dominance."[3] Nancy M. Henley sees touching in business as a sexist powerplay, a subtle act of downsmanship that keeps woman in her lowlier place. Men may touch women far more than women may touch men; superiors (usually men) may touch subordinates, but subordinates may not touch superiors; older (usually men) may touch younger. The politics of touch maintains the status quo, reinforcing the male-dominated hierarchy.[4]

In a few situations, however, these unspoken rules are acceptably breached. Any one may touch any one else when congratulating for spectacular accomplishment, offering encouragement, or giving condolences.

The president of a corporation will not feel his receptionist is out of bounds if, while patting his arm or shoulder, she is exclaiming, "Wow! Congratulations! I heard last night that your son won an Olympic gold medal." At that instant he is primarily a proud father and secondarily a corporation president.

A teaching assistant holding the lowliest position in the pecking order of a university department could acceptably touch its chairman once a year when he gave his Faculty Lecture. As he left to deliver his talk, about which he was always nervous, she would give him a "good luck" pat on the shoulder.

A friend who has worked fifteen years in a multinational organization remembers touching the chairman of the board just once. "His son had committed suicide. I ran into him in the hallway. It seemed incomplete just to express my regrets in words. I gripped his forearm while we talked. We were parents more than employer and employee, commisserating with each other on the sorrows that can afflict parents."

Inevitably business will have to draw up a Code of Ethics about touch and sex. Margaret Mead said, "If we're going to have men and women in business on an equal basis, with men over women and women over men, we have to develop decent sex mores. It is very difficult to run an army if the general is in love with the sergeant."

I have these suggestions for such a code:

1. That it be drawn up using the growing principle of "codeterminism" in which all segments of an organization's population—males, females, old, young, potwashers to presidents—are represented. Students are now on the Boards of Regents of universities; consumers are on the boards of corporations. Since ethnic attitudes toward touch vary, ethnic groups in the work force should be represented among those writing the code.
2. That it use these clarifying definitions of touch. *Person-Centered Touch* is tactile contact that *gives* to the recipient support, encouragement, congratulations, admiration, affection, thanks, appreciation. *Body-Centered Touch* is that which *demands from* the recipient sexual favors, reassurance of one's worth and desirability, extraordinary hours, places or conditions of work, or

that makes any presumption of intimacy unwelcome to the receiver.

3. That, to protect personal rights, touching in business be considered acceptable if the two people both want the contact and there is no coercion; if they are discreet; and if their behavior does not encroach upon others' schedules or interfere with their job responsibilities and the company's productivity. If two employees who are in love want to grab a kiss or enjoy a pat on the behind when they are alone in the elevator, company rules ought not to deprive them of this pleasure.

People who work in "folk hero careers" have unique ethics and mores of touch. Workers in these careers—show business, designing, modeling, sports, politics—are in the public eye, their names and faces known to many. In this era of "people journalism," we have almost a day-to-day familiarity with how they spend their time, what parties they attend, where they eat, shop and have their hair done, how much they earn, and so on. Built into the easy camaraderie of such careers as acting, designing and modeling, which may involve hurried changes of clothing in the presence of other people, there is a casual, sometimes good-natured acceptance of the tap, pinch, hug or feel. Some actors and models of both genders are irritated by these bodily contacts, but accept them because "it comes with the territory"; those who protest risk being thought of as unfriendly grouches. As Harry Truman said, "If you can't stand the heat, stay out of the kitchen." The Screen Actors' Guild has a "Morals Complaint Board" to investigate charges from those who refuse to be casual about touch and sexual overtures.

From *performer-celebrities*, we expect uninhibited and daring behavior. Their free-spirited actions often endear them to us; they act out that part of many of us that would like, but doesn't dare, to be equally uninhibited. But from our *politi-*

cian-celebrities we expect behavior that is above reproach. Let a male candidate for public office hug a female campaign worker or stroke the arm of a male follower just a fraction too long, and the rumor mills start churning. Is he a lecher? Or gay? Can we trust a person like that to act honorably if we elect him?

We might be pleasantly titillated by the mild flirtation between Barbara Walters and Cuban Premier Fidel Castro in a television special, because it livens up our lives. But if we observed this same kind of flirtation between Castro and an American woman politician, we would respond differently. We would wonder what their attraction for each other might conceivably hold for us and for the nation.

Why do we care more about the personal behavior of politicians than that of other folk heroes? For several reasons. One, because we elect politicians and give them their power, so we have an ego-investment in how they behave. If they behave with propriety, they vindicate our judgment and good sense in having selected them. If they stray from standards that we consider proper, we are upset, because they make us doubt our ability to assess character. Our relationship with those we elect often carries a hidden "balance sheet"—"We gave you power, now you owe us 'proper' behavior."

We also care about politicians' personal behavior because their didoes might wind up costing *us*, in the form of wasteful spending of public funds. Should a politician take a sweetheart on tax-deductible trips as an aide or secretary, paying her salary and expenses out of public funds, we feel, How dare he enjoy himself on our hard-earned money? We may, in fact, care less about the morals than the money. If he were to take the same trip with the same woman but pay for her out of his own pocket, many of us would not be upset.

Politicians and constituents have confused psychological relationships. They are both parents and children to each other. When a politician grasps your hand, giving it a warm hearty shake or a comforting squeeze, he is a "nurturing parent" whose touch is saying, "I will take care of you. You can count on me to help you get your needs and wants filled. Just vote for me." James C. Davies, a political scientist specializing in studies of human nature in politics, says that we choose the candidate who we believe will best help fill our psychological and physiological needs. It is the wishful-believing-child part of us that marks the ballot or pulls the lever on the voting machine for the Daddy we think will make our hopes come true.

At the same time we are "parent" to the politician, saying in effect, "You behave, son, otherwise I'll be disappointed in you. Do me proud by always making decisions of which I approve and behaving like a fine, upstanding, moral man." The child part of us puts him into office but the parent part of us gives him gold stars or black marks for how he behaves there. Politicians know very well that they need to walk a tightrope of acceptable behavior.

To uphold their images, and to establish a happy balance between being personal and friendly and yet not being too intimately physical, politicians have worked out a touch ritual unique to them. Observing people in various professions, I could find nobody except politicians who frequently practice this gesture. They shake hands with the right hand while simultaneously patting with the left hand the side of the waist of the one being greeted. This enables an officeholder to enfold a constituent within two hands—a kind of diminished hug. Watching on television the hostages' homecoming from Iran, I was mesmerized by George Bush's well-orchestrated hands rhythmically dancing this "political pat." As each hostage passed in the receiving line, Bush shook the individual's hand with his right hand and waist-patted with his left. Try this yourself; it is no small feat of dexterity, pumping up and down with one hand while pumping sideways with the other.

Waist-patting could also be interpreted as an expression of

impatience, a subliminal shove saying, "Come on, move along, so I can get this damned task over with." Standing in receiving lines is an occupational hazard that politicians dislike but endure. A receiving line sometimes comes up with a shock. Patricia Nixon, touring a foreign land, was startled to have an arm thrust at her but it had no hand; she graciously shook the stump. An office seeker remembers how "I turned hot and cold and started to shake" when a one-legged Vietnam veteran sitting in a wheelchair playfully held out his empty pant leg for the candidate to shake. Gerald Ford, shaking hands down a line of college students, was startled when a stranger said, "Hi Dad. I'm your son." The young man had grown a beard since his father had last seen him.

In the worlds of business and politics the most powerful of all tactile contacts is the handshake. Knowingly or not, many of us base our judgment of another's character on how that person feels to us when we shake hands. Handshakes can be warm, cold, hearty, firm, flabby or moist. "A handshake," says international banker David Rockefeller, "helps determine what dam gets built and who gets a job." Another banker says, "A man or woman who gives you a handshake that feels as if you're touching a wet noodle or a mound of pudding is not going to inspire confidence in their ability to administer a million-dollar loan. A man whose handshake is so powerful that it makes you wince makes you wonder if he is insensitive in other ways as well."

Shaking hands, we are engaging in palm talk. Our palms may be saying, "We are peers. . . . We acknowledge that we have access to each other for business dealings. Here is a promise of a continuing relationship and future contact." Our palms could be saying, "I do this as a greeting and farewell ritual. Don't give it any more meaning than that." Or, if the handclasp is prolonged, our palms may be hinting that "I'd like to know you more intimately." Counseling women who have executive-level jobs, I recommend that at the start of a conference, they initiate handshakes with one or two men, especially those whom they do not normally see in the course of a day's work, as a symbolic gesture to establish their peer status with the males present.

Whether it be in the worlds of business or of politics, there is one kind of touch about which no one ever complains. In fact, as surveys of job satisfaction show, nearly all working people complain that they don't get enough of it. This "least sexually loaded touch"[5] is welcomed by all. It is a pat on the back, recognition for doing well.

Let us celebrate our ability to give pats on the back. As robots replace humans in the workplace, the only remaining distinction separating man from monster may be our ability to "press the flesh"—warm, live, breathing flesh. No robot, perfected though it may be, can ever achieve that.

13

Touching in Other Countries

If you were suddenly to come upon a strange woman in her bath in various parts of the world, here's what she would probably do.

A Mohammedan woman would cover her face.
A Laotian woman would cover her breasts.
A Chinese woman (before the Revolution) would hide her feet.
In Sumatra the woman would conceal her knees.
In Samoa she would cover her navel.
In the Western world she would cover her genital area.

Such a variety of reactions to being seen nude indicate the variety of responses people have throughout the world to touching and being touched. Each of the world's 166 nations has evolved its unique manners and mores about the appropriateness of one human being making contact with the skin of another. For those of us who want to express friendly feelings toward peoples of other lands by touching them, matters are complicated by the absolute absence of a pattern to guide us as to when, where, with whom, with what parts of the body, we may acceptably do so. National touch codes are full of what seem, to Americans at least, to be inconsistencies and anomalies.

We would expect religious countries like Italy and Greece, in which the church has controlled sexual behavior and bodily intimacies, to be more reserved physically. The opposite is true. Greeks and Italians are among the most touching of all the world's people. A stranger, entering a Greek home, may be welcomed with a warm embrace and kisses on the cheek. In Roman Catholic Italy, it is common for men on the street to pat or pinch the behinds of strange women they admire. Rather than feeling flattered, women resent it. Foreign women see it as evidence of great macho-ism on the part of Italian men and of the need for their Italian sisters to be liberated, especially from the insulting gesture of having one's crotch grabbed by a strange man. Actress Elizabeth Taylor has told of walking out of the Roman Coliseum with her behind "black and blue" from the pinches of Italian men.

We would expect Sweden, with its liberal attitudes toward sexuality, to be a highly tactile culture. It is not. Brigham Young University has prepared, in conjunction with religious and missionary groups, travel agencies and international consulates and diplomats, a series of *Culturgrams* which briefly describe customs, manners and life styles of more than 75 cultures. The Sweden *Culturgram* advises that "such gestures as embracing or putting one's arm around the back of another are unusual, unless among very good friends." (Culturgrams, Brigham Young University, Center for International and Area Studies, Box 61 FOB, Provo, Utah 84602 (801) 378-6528.)

Further pointing up the inconsistencies of touch, some countries that have strong taboos against homosexuality—the Arab nations, Italy, Greece, Korea—are countries in which it is acceptable for males to put their arms around each other and hug in public, to walk arm in arm or hand in hand down the street, even—in close relationships—to kiss each other on the mouth. Egyptian President Anwar Sadat, sitting and talking with another male, often put a hand on the man's knee. Several years ago a group of Italian-Americans in New York City organized Brotherhood in Action, a forum at which they could share the problems they were experiencing adjusting to life in the United States. One man told of the shame he was made to feel by his new American friends because as a teenager he kissed his father on the mouth.

Learning that the same touch has exactly opposite meanings in different lands adds to our confusion. If you were talking to Fiji Islanders and you folded your arms in a "hug yourself" gesture, they would be pleased. You would be showing respect for them. Do the same thing while you are talking to Finnish people and they might consider you arrogant. In Finland folded arms are a sign of disrespect, a distancing gesture that says, "I am the high and mighty one, superior to you."

If you were to caress the head of an adorable native child in Singapore, Thailand or Taiwan, the parents might be upset with you. They consider the head to be sacred. If you were in Mexico and caressed an adorable child with the same kind of touch, her parents might smile gratefully. The Mexicans have a saying, "A child that is not touched will be unlucky."

A style of touching that is common practice in one land is unacceptable in another. An Argentine male often fingers another male's jacket lapel while talking to him. His touch might be almost a caress, saying, "I am feeling friendly and warm toward you." An Australian male, whose idea of acceptable body space between men is very different, would choose not to get close enough to another male to finger his lapel, except perhaps in a hostile situation, when his touch would be a combative one saying, "Listen, one more move or word out of you and I'll let you have it."

The *Culturgram* on Australia says, "Outward signs of emotion, such as hugging, are not considered manly and are avoided."

We can make no assumptions that members of the same ethnic group or the same region will react alike. We cannot say, "All Latins accept this," or "All Nordics do that," or "All Middle Easterners have this touch custom." I was shocked when I first saw people picking their teeth in public in Rome, until Italians told me this is a widely practiced custom, sometimes considered a sign of admiration for the cuisine in a restaurant. Indeed, it is a status symbol for an Italian male to wear a gold toothpick dangling from his tie or watch chain, to be detached for use in public whenever particles of food in his teeth so move him. Other Latins, particularly Spaniards and Uruguayans, consider this self-grooming touch the height of bad manners.

Is it any wonder people around the world have so much trouble understanding each other? We have been learning about each other's politics, economic systems, religions, gross national products, native cuisines, but not about one another's sensitivities. We know more about how nations eat than about how they touch. In our quest for worldwide fraternity, we have been engaging in international relations with diminished faculties. One of our senses—our sense of touch—has been missing.

Cultural anthropologist Edward T. Hall believes that our lack of knowledge about one another's acceptable body closeness and ways of touching cost us a high price in global tensions, prejudices, ineffective business dealings, and distaste for one another's ethnicity. We are antagonized by others before we even know them. During a long and distinguished career, Hall has been studying the attitudes of different people toward touching, body space between individuals and concepts of time. He has coined the word *proxemics*, derived from *proximity*, to describe the behavior he studies. How much proximity are we willing to endure? What kinds of touches are we comfortable giving and receiving when we are in proximity to one another? Intercultural antagonisms originate when we feel that another person's body may be too close or too far from us.

When a Japanese bows from the waist in acknowledging an introduction to you, rather than shaking hands as Americans are used to doing, you may feel put off by his "keep your distance" message. To shake your hand the Japanese would have to be in closer proximity; his body space would be invaded. Many of us misinterpret the customary body space of the Japanese as a sign of unfriendliness. Later in this chapter we shall see why they do it. Similarly, when Mexican businessmen come to negotiate deals in the United States, they may be offended by the distance of executives who sit behind desks. In the Mexicans' own offices they keep sofas on which they can sit close, touch and gesticulate while negotiating. After polite preliminaries, a Mexican businessman is apt to rise from his desk, take the visitor by the shoulder or arm and propel him toward the sofa, saying, "Come, let's talk."

If a "territorial American or German" who likes a large space around his body is visiting or attempting to do business with Middle Eastern, Latin American or Southwest Asian businessmen who are used to a smaller body bubble space, he feels stress. He tightens his muscles, holds in his breath, feels his heart palpitating and irritably wishes the "pushy foreigner" would stop "breathing down my neck," "crowding me," "sticking his face into mine," or "spraying me with his saliva." The foreigner's manner may not really be pushy; it may even be polite and deferential. But the American equates the person's manners with "pushiness." The foreigner doesn't know any other way to behave; he has been doing that all his life. To him, the visitor is a cold and aloof foreigner.

A major problem between Arabs and Westerners is that we simply do not understand that Arabs thrive on close bodily contact. Arabs have high "sensory involvement" with one another. Edward T. Hall says:

> The Mediterranean Arabs belong to a touch culture, and in conversation they literally envelop the other person. They hold his hand, look into his eye and they bathe him in their breath. The American on the receiving end can't identify all the sources of his discomfort, but he feels that the Arab is pushy. The Arab comes close, the American backs up. The Arab follows, because he can only interact at certain distances. Once the American learns that Arabs handle space differently and that breathing on people is a form of communication, the situation can sometimes be redefined so the American relaxes.[1]

I have experienced exactly what Hall describes. I once briefly shared a vacation cottage with an Arab many years my junior, because one of us had arrived too early for our rental period. One evening we stood talking in the kitchen. I was leaning against the counter. The young Arab stood at right angles to me so that his upper arm and shoulder occasionally touched mine as he talked with intensity. I backed off. He moved forward. I behaved the way my American culture had conditioned me to act. I gave his touching sexual meaning, became annoyed, and ended the interesting conversation which I had been greatly enjoying. How I wish I had read Edward Hall's books at that time! Now I realize the young man was behaving in the manner in which he had been conditioned by his Arab culture. That incident was one of the life experiences that inspired me, years later, to write this book. (I recommend Edward Hall's fascinating books, *The Silent Language*, *Beyond Culture*, and *The Hidden Dimension*, because they are clarifying and fun, and they wind up making you anxious to meet foreigners so you can test his observations.)

Hall defines the space between two bodies in four dimensions: *intimate, personal, social* and *public* distances. Of course there is some overlapping of spatial distances on some occasions. At a cocktail party we may move quickly through intimate, personal and social spaces.

Intimate distance may be anywhere from no space between bodies, as in lovemaking, up to eighteen inches. All of one's body is accessible for touching in intimate distance, either with our hands or with other parts of the body. We feel each other's "thermal" touch by the radiant heat our bodies give off. (The one activity besides lovemaking in which the most intimate distance applies is wrestling!)

Personal distance, from one and a half to four feet, is the distance between two people in which touch is always potentially possible. Two people having the farthest reach of personal distance between them, four feet, can still "touch fingers if they extend both arms,"[2] and that puts them within each other's sphere of touch. Whether or not they actually touch, they are still within a physical space in which touch is possible. Americans spend much time in personal-distance space with spouses, children, relatives and close friends. Although we may not do much touching, that amount of bodily bubble space is comfortable for us.

Social distance, four to seven feet, is customary when we are with individuals outside our personal relationships. That much space makes it awkward or impossible to reach out and touch. Social distance marks the space between you and the person who may be interviewing you for a job, you and the salesperson who is writing up your sales check, you and the waiter who is taking your order, you and the professor to whom you are talking about your grades. I just measured the space between me and a counseling client. I sit in front of my desk, not behind it, and the client sits on a sofa. We are about fifty inches apart—a social distance. I keep a low chair next to my desk. Should I want to comfort a client, I pull up the chair near her so I can take her hand, or I move over and sit next to her on the sofa, going from social distance to intimate distance for my touch of comfort.

Public distance, anywhere from twelve to twenty-five feet, is that reserved for public officials, theatrical stars, public speakers and for public occasions. The potential for touching does not exist in public distance, although it is often momentarily breached by someone who runs up to touch a celebrity (thus going into intimate distance) so he can boast afterward that "I was as close to her as I am to you. I was close enough to touch her."

Americans act out intimate space mostly in sexual relationships beginning with seductive postures and ploys. A man sits on a sofa with a woman and moves close enough to put his arm around her shoulders, with no space between their torsos. In some countries people get that close to you as part of their social distance; they intend no sexual meaning. Features of American intimate distance are present in Russian social distance. A Russian man will sit that close to another man and put a comradely arm around his shoulder. An American male whose space is thus invaded would feel uncomfortable.

Given the fact that every human is born with the need for touch, pressure and heat from another human body, why have different lands developed such varying attitudes toward such a universal need? What factors influence how people feel about touching themselves and others? Population density, climate, politics and religion, sanitation, superstition and mythology, tradition, social status, attitudes toward women, and attitudes toward the body all affect touch behaviors.

Population density. Intense overcrowding makes people either accepting or rejecting of bodily closeness, especially between strangers. How well people are able to cope with the jam-packed spaces of population density often depends on how they feel about being touched by strangers. The males of Cairo, Beirut, Damascus and other Middle Eastern cities are casual about riding on buses and trolleys and having their buttocks, thighs, legs, torsos intimately pushed against or jostled by strangers. Two or three men stand on the same step holding on to the same handle as they dangle outside a trolley; it is only by pressing against one another that they are able to maintain their precarious hold. This helps us to understand why Middle Eastern men are comfortable being close to foreigners; closeness is often a survival mechanism for them.

Population density has the opposite effect in Japan. There the people have a saying, "Touching without feeling," which enables them to handle their unease about being packed into public conveyances pressing strange bodies. They avert their eyes, wear a blank public mask and look into the distance; they do not make eye contact with those with whom they are in intimate bodily contact (exactly my behavior when I used to ride jammed New York subways).

In such crowded conditions females may find themselves the target of groping, probing anonymous hands. A pawed native-born woman will usually remain passive and say nothing as she tries to edge away, but outspoken foreign women are apt to protest. An American woman carries an open safety pin to jab at hands "snaking toward my crotch." A popular story in Tokyo tells of the American resident who was touched familiarly every day as she rode to work with many of the same people in the commuter train. One day, feeling fingers creeping on her thigh she reached down, seized the hand, jerked it above the heads of the crowd and loudly demanded in Japanese: *"Dare no te desuka?"* ("Whose hand is this?") The hand was attached to an elderly man in a business suit; he slunk away. She was never again bothered on her daily ride to work.

I can understand why the Japanese continue their long tradition of bowing instead of shaking hands in greeting. When you are so often tightly packed in with other humans, it must be a relief to keep your bodily distance whenever you can. Tourists in Japan often feel not only culture shock but acute physical symptoms as their bodies adjust to hordes of humanity in a land that has more than half the population of the United States—about 115 million—in only 4 percent of the land area.

Climate. When the weather is cold enough people get close, no matter what their "proxemics" may be. Body heat is too precious a source of comfort to let tradition stand in the way of enjoying it. Edward Hall tells of a colleague who was caught in a snowstorm while traveling with companions in the mountains of Lebanon. They stopped at the nearest house and asked for shelter overnight. The house had only one room. Instead of putting the strangers on the other side of the room, the host placed them next to the pallet where he slept with his wife, "so close their bodies almost touched."[3]

The Japanese who maintain bodily distance with foreigners may behave differently when the weather is freezing. An American priest explained:

> To really know the Japanese you have to have spent some cold winter evenings snuggled together around the hibachi. Everybody sits together. A common quilt covers not only the hibachi but everyone's lap as well. In this way the heat is held in. It's when your hands touch and you feel the warmth of their bodies and everyone feels together—that's when you get to know the Japanese. That is the real Japan![4]

Politics and Religion. The Spanish and the Chinese have been at opposite ends of the political spectrum, the former dominated by the Catholic Church and having had for many years a fascist government, the latter with little religion and having a Communist government. Yet both cultures have, in the past, frowned upon men and women touching affectionately in public. During Franco's rule, a college couple were caught kissing while strolling on the grounds of the University of Madrid. They were arrested, threatened with expulsion and severely reprimanded. Visitors to China today do not often see affectionate touching between men and women, although they do see it between the elderly and small children. Eleanor and Sam Katzman, Los Angeles marriage counselors, reported after a trip in which they studied marriage, divorce and sex customs that in only one Chinese city, Shanghai, did they see couples hold hands or put their arms around each other. However, this is changing, as Chinese and Western cultures are exchanging tourism and commerce.

The Spanish discourage touch because they feel that it might cause people to have erotic feelings and lead to temptations of the flesh. The Chinese discourage bodily contact because they believe that people's energies must go first and foremost into building the socialist state. To have one's mind on romantic aspects of life would deflect people from their ideological goals.

Sanitation. In India, parts of the Middle East and parts of Africa, you do not touch with your left hand, eat with your left hand, pass an item to another with your left hand. Columnist Jimmy Breslin interviewed a black man who tried resettling in Tanzania and then returned to the United States. "They never told me," said the man, "but when you use your left hand to pass something to somebody it's considered an insult. People believe that the left hand should be used in the bathroom, and therefore anytime you reach out to anybody, you use the right hand. So here's me and my family from Harlem walking around with our left hands out and we had the whole of Tanzania mad at us." Some day, hopefully, income levels and national priorities will rise, so that people everywhere will have hot and cold running water—a sanitation facility that Americans take for granted. There would then be plenty of water with which to wash after toileting, and this touch taboo might slowly fade away.

Superstition and mythology. A Greek myth says a stranger might be a god coming down to earth in disguise; that's why, as mentioned earlier, the Greeks often greet strangers with hugs. (Culturgram, Greece, Language and Intercultural Research Center; Brigham Young University, Provo, Utah 84602.) Superstitions either encourage or inhibit the touching of substances. A touch superstition common to both Israel and Ireland has people knocking on wood after someone is praised or enjoys a stroke of luck or is experiencing good health. Touching wood is supposed either to scare away evil spirits who are said to live in trees or to invoke the protection of trees thought to be sacred. The Baduis, a Javanese tribe that isolates itself from modern civilization, forbids its members to touch metal because "metal can hurt the earth." (As a teenager I was forbidden to touch stainless-steel pots while I was menstruating, because the "acid" in my body would turn the pots black!)

Tradition. In prerevolutionary China it was acceptable for government officials to engage in a touch gesture that Americans consider more fitting for street toughs. Once, when Madame Chiang Kai-shek was visiting the White House, she and Eleanor Roosevelt discussed the right of the press to criticize officials. Indicating how she would deal with the press if it dared criticize her, the delicate, doll-like Chinese lady "raised a manicured index finger and drew it slowly across her own throat."[5] Ms. Roosevelt was shocked by the gesture, one that had been traditional in Madame Chiang's own land.

Social status. In lands ruled by monarchies people have been raised with rigid taboos against touching those in high positions. A Dutch writer reports this incident:

> Years ago the Queen of Holland took her daughter, then eight years old, to the zoo. The little girl saw an animal that caught her interest and left her mother for a moment to look in the cage. Another visitor recognized the princess and stroked her hair. The Queen was furious, went over to the man and said, "You shouldn't touch a

> princess of the House of Orange." The touch of a common citizen could rob the princess of her dignity—could degrade her magically.[6]

Citizens might be discouraged from touching a child of a presidential family in the United States, but that would have more to do with the child's safety, not because she would be degraded by a touch.

Attitudes toward women. One aspect of researching and writing this chapter saddens me. I had never before even thought about this subtle discrimination against women in some parts of the world—just one more of the ways in which women lead deprived lives. They are never touched by strangers, and their husbands may touch them mostly for sex. Reflecting women's subordinate situation, there is little concern that they too need cuddling, stroking, caressing and pleasurable stimulation through their skin. Entering a home in many countries, a male visitor shakes hands with the man and nods to the wife, but is expected never to offer her a courtesy handshake. In rural areas of Moslem and Hindu countries and in orthodox households in large cities, women are never even seen by male visitors. They prepare the meal, but it is served by young males in the family or by male servants. "Sometimes," an American businessman told me, "you see a pair of female arms thrust through a door handing out a dish but never any more than that."

The taboo against strange males touching women is so powerful that a male doctor in rural areas may not be permitted to perform surgery on women. A husband-and-wife team of British doctors told writer Agatha Christie how they performed operations in Iraq: "It was impossible for a Moslem woman to be operated on by a man. He was not allowed to see or touch the patients. Screens had to be rigged up. The husband would stand outside the screen with his wife inside; she would describe to him the conditions of the organs as she arrived at them and he, in turn, would direct her how to proceed."[7]

It is in their kinship with other women that many of the world's females get touched. Women in other countries embrace, kiss, shake hands, or clasp arms in greetings and goodbyes much more than American women do. But other than that, they are looked upon as nonpersons unworthy of being touched by foreign visitors. This makes me impatient for equality to become an accomplished fact. I would like to incite the world's women to start a Handshake Revolution in which we initiate touch of one another by thrusting out our hands in greeting, even when a hand has not been offered to us. Such hand-to-hand gestures could become symbolic of women's burgeoning freedom. I would feel especially good if I thought my initiating a handshake with a woman, and with a man in the presence of his wife, would make her feel worthier of offering a handshake herself, a little less willing to accept her unequal status. She may titter or giggle in momentary embarrassment, but she would get my message: "You are a woman who is entitled to do as I do."

Attitudes toward the body. In some countries it is acceptable if you have an anal or genital itch in public to touch yourself on the offending spot and scratch through your clothing. Americans consider such self-touching in public to be vulgar; we are uncomfortable with lusty, earthy, naturalistic attitudes toward body functions and smells. Italians and Arabs are especially accepting of the body. Italian women, for instance, do not shave off underarm hair. Their attitude is that "that's part of the body; it's nothing to feel ashamed of or to consider ugly." In the Middle East the spices and seasonings people add to food give off pungent mouth odors. People believe that "You don't disguise these natural odors with deodorants and mouthwashes."

As I walk and drive through the streets of Los Angeles with its rapidly growing, various ethnic populations, I am becoming aware of the self-touching habits of newcomers to our land. I am less and less shocked when I see a man standing at a corner waiting for the bus scratching an itch in his private parts. If you have ever had an anal itch crying out for instant care, you may agree that there is something to be said for those cultures which go right to it. They must find their instant self-touching far more relieving than we do our ploys of finding a secluded doorway or surreptitiously scratching behind a newspaper or a handbag. As people with varied ethnic backgrounds become assimilated into the American culture, they will probably wind up a bit more inhibited, and we may wind up a bit less inhibited, about what we all do with our bodies in public.

Underlying our difficulties in understanding and accepting other people's customs regarding touches and physical closeness are two phobias. One is *xenophobia*, "fear of strangers," which is as old as the human race. The other phobia we increasingly experience as the world becomes a global village. No word exists for it and so I am coining a word, *ethniphobia*, "fear or dislike of ethnic differentnesses; distaste for the ethnic customs and gestures of others." Samenesses—traits, customs and touch gestures to which we have been habituated—are familiar, safe, secure, comforting. Differentnesses mean we have to undergo an inner adjustment to new behavior. Our ethniphobic mind-set makes us automatically think, How crude. How uncouth. How distasteful. How bizarre.

But we can choose not to be ethniphobic by changing our mind-set so that when we observe strange and unfamiliar behavior we automatically think, How interesting. I'm learning something new. Humans are fascinating in their infinite variety. Liv Ullmann toured famine-stricken lands on behalf of UNICEF. In Bangladesh she touched and held hands with a woman. When Liv left, she gave the woman a hug. She felt the woman drawing away. Through her interpreter, Liv asked why. The woman answered, "In my country we kiss feet when we say goodbye." Liv, a quintessential lady, bent down and kissed the woman's feet. Then they hugged—each woman having exchanged the parting ritual of her own world.

This chapter has been useful in ridding me of *my* ethniphobia. I have done little foreign travel; I plan to do a lot more in the future. I have determined to look upon differentnesses with interest, not distaste or scorn. Julie Nixon Eisenhower has written of a differentness in China that I look forward to observing. "I discovered that spitting was a socially acceptable custom when we met with a high official and right in the middle of a sentence, he cleared his throat and used the spittoon between my chair and his."[8] I would have been aghast at this "differentness." Now I won't be. Instead, I plan to observe the expertise of the spitter's aim!

While touring China, Michigan Governor William Milliken had dinner with Vice Premier Deng Xiaoping. "He kept a large spittoon between us," reported Milliken, "and every few minutes he'd muster all his strength and all the noises that come with it and spit into the bucket. Every time he did, I moved out of the way."

Visiting other lands, I intend to ignore some existing mores and initiate new ones. Carol Saltzman, associate dean of UCLA's Office of International Students and Scholars, is in charge of research into touching between cultures. "How touching is perceived," she says, "depends upon the context in which it is done and the manner."

I shall touch man, woman and child gently, with light pressure on a nonvulnerable, nonthreatening part of the body, forearm, hand, or shoulder. I shall touch in thank-you, in friendship, in compassion, in joyful acknowledgment of a pleasurable occasion enjoyed together. I shall be sensitive to other people's rights and privacy. My manner and my smile will make it clear that my touch speaks of friendship, not disrespect. I may be taking a chance that a Britisher may call me "cheeky"; a German might call me "pushy"; a Japanese might consider me "bold"; a Spaniard might call me "sacrilegious." The fact that we are all born with a need for touching will take precedence for me over the acculturation each of us has received.

Even as you read these words, touch codes are changing because of several factors: men and women are doing business as peers; the younger generation is attending schools; affluent families are vacationing—in one another's lands. All over the world men accustomed to doing business only with men now find themselves shaking hands with, and accepting closing-of-business-deals pats and embraces from, women executives. No matter how sexist a man may be in his personal life, he gladly exchanges congratulatory pats, handshakes and hugs with female executives from whose companies he hopes to gain profits. Lester B. Korn, head of an international executive-search firm, says that the successful executives of the future will be those who have learned "to deal with different cultures."

Increased commerce is always followed by increased tourism. By the year 2000, tourism will be one of the world's biggest industries, if not *the* biggest. Two billion people are expected to be exploring other lands every year. Norman Cousins urges us not to "let governments be the sole custodians of human destiny." Let us all help to create our own destinies by lovingly touching around the world, simultaneously crossing the boundaries of all kinds of skins and all kinds of states.

14

Homo Biologicus

At the moment when I write this, and at the moment when you read it, somewhere in the world someone is doing research on the most remarkable machinery we will ever know—the human body and brain. Nothing we create will ever surpass that exquisite design—us.

Since our beginnings, we have been acquiring knowledge and wisdom. We have been *Homo sapiens*, "Knowing human." Now we have enough knowledge to know that we must enter our next dimension, *Homo biologicus*, "Biological human."

Some of the human race's *angst* has been caused by the fact that, as Arthur E. Morgan, former president of Antioch College, put it: "Our biological drives are millions of years older than our intelligence." Our biology says, "I need this to feel good." Our sociology says, "No, manmade rules, ignorance and superstition forbid you to have it."

Learning more about our biology, we need to think and reevaluate ideas on which the human race has based its behavior thus far. We must start thinking from a new basis—Biological Human. What does our biological organism require to function at its optimum, its most joyful? We now have scientific knowledge that among its powerful needs is human contact. Dr. James Lynch says, "There is a biological basis for our need to form human relationships. If we fail to fulfill that need, our health is in peril."[1]

Those who combine *Homo sapiens* and *Homo biologicus* will be the pioneers of the New Human Being—one who lives and acts in accordance with knowledge available to us as we approach the twenty-first century. The New Human Being will be the first to take that giant step across the chasm between our biology and our sociology.

Often making contact with another, the New Human Being will add to the health and joys of the world.

NOTES

Chapter 1 (pages 1–20)

1. Rod McKuen, *Alone* (New York: Pocket Books, 1975), p. 11.
2. W. Heron, "The Pathology of Boredom," *Scientific American*, January 1957, p. 52.
3. Richard Gill, *White Water and Black Magic* (New York: Holt Publishing Co., 1940), pp. 208–09.
4. *Today*, NBC-TV, Jan. 19, 1978.
5. Somerset Maugham, *The Summing Up* (New York: Doubleday, Doran, 1938), p. 78.
6. Desmond Morris, *Manwatching: A Field Guide to Human Behavior* (New York: Harry N. Abrams, 1977), p. 266.
7. Frank A. Geldard, "Body English," *Psychology Today*, December 1968, p. 44.
8. "The Anatomy of the Brain," *Time* magazine, Jan. 14, 1974, p. 55.
9. W. Penfield and T. Rasmussen, *The Cerebral Cortex of Man* (New York: Macmillan, 1950), p. 214.
10. Liv Ullmann, *Changing* (New York: Knopf, 1977), p. 12.
11. Meta Carpenter Wilde and Orin Borsten, *A Loving Gentleman* (New York: Simon and Schuster, 1976), p. 138.

12. Desmond Morris, *Intimate Behaviour* (New York: Bantam Books, 1973), p. 107.
13. Edward T. Hall, *The Hidden Dimension* (Garden City, NY: Anchor/Doubleday, 1969), p. 62.
14. Phyllis Barber, "Touching," *Utah Holiday Magazine*, June 1980, p. 27.
15. "Open Forum," *American Civil Liberties Union Newsletter*, Los Angeles, July–August 1978, p. 1.

Chapter 2 (pages 21–37)

1. F.B. Dresslar, "The Psychology of Touch," *American Journal of Psychology*, vol. 6 (1894), pp. 313–68.
2. Annual Conference of the Osbstetrical and Gynecological Assembly of Southern California, Beverly Hills, April 1975.
3. Hannah Tillich, *From Time to Time* (New York: Stein and Day, 1973), p. 25.
4. René Spitz, *The First Year of Life* (New York: International Universities Press, 1965).
5. Jerry L. White and Richard C. LaBarba, "Effects of Tactile and Kinesthetic Stimulation on Neonatal Development," *Psychological Abstracts* no. 12501, 1977.
6. Mary McFall Jankovic, "A Study of the Physiological and Behavioral Responses of Low Birth Weight Neonates Who Have Been Held and Stroked During Feeding," *Emory University Research Abstracts*, 1973–74, p. 11.
7. Thomas J. Boll, Stanley Berent and Herbert Richards, "Tactile Perceptual Functioning as a Factor in General Psychological Abilities," *Perceptual and Motor Skills*, April 1977, pp. 535–39.
8. Help for Brain-Injured Children Foundation, 981 N. Euclid, La Habra, CA 90631.
9. Mother and Child Enrichment Group, 6546 Hollywood Blvd., Suite 201-M12, Hollywood, CA 90028.
10. Liv Ullmann, *Changing* (New York: Knopf, 1977), p. 230.
11. James W. Prescott, Jr., "Body Pleasure and the Origins of Violence," *The Futurist*, April 1975, pp. 64–73.
12. Selma Fraiberg, *Every Child's Birthright* (New York: Basic Books, 1977), p. 62.

13. Helen A. DeRosis, "Violence, Where Does It Begin?," *The Family Coordinator*, October 1971, p. 361.
14. *Today*, NBC-TV, Feb. 28, 1979.
15. Dr. Samuel Woodard, Howard University, Washington, DC, personal conversation with the author, Feb. 1, 1982.
16. Jane van Lawick-Goodall, *In the Shadow of Man* (New York: Houghton Mifflin, 1971), p. 237.
17. Marvin E. Wolfgang, "Violence and Human Behavior," in *Proceedings of the Annual Conference of the American Psychological Association*, Washington, DC, Aug. 30, 1969.
18. *UCLA Monthly*, Alumni Association, March–April 1981, p. 1.
19. "What Makes Me a Unique Being?" *Newsweek*, June 21, 1971, p. 66.

Chapter 3 (pages 38–59)

1. Robert Coles, M.D., "Touching and Being Touched," *The Dial* (Public Broadcasting Corporation publication), December 1980, pp. 26–30.
2. *Lansing State Journal* (Michigan), Jan. 9, 1981.
3. Vidal S. Clay, "The Effect of Culture on Mother-Child Tactile Communication," *The Family Coordinator*, July 1968, pp. 204–10.
4. "Substitute Bonds—Drugs as 'Family,' " *Brain/Mind Bulletin*, (P.O. Box 42211, Los Angeles, CA 90042), vol. 5, no. 12 (May 5, 1980), p. 3.
5. Arthur Janov, "For Control, Cults Must Ease the Most Profound Pains," *Los Angeles Times*, Dec. 10, 1978, part 6, p. 3.
6. "Dear Abby," University Press Syndicate, (4400 Johnson Drive, Fairway, KS 66205), August 1, 1969.
7. Manfred Clynes, "Sentic Cycles, the Seven Passions at Your Fingertips," *Psychology Today*, May 1972, p. 59.
8. Sidney B. Simon, *Caring, Feeling, Touching* (Niles, IL: Argus Communications, 1976), p. 100.
9. William Hamilton, "The NOW Society," March 12, 1974.

Chapter 4 (pages 55–65)

1. William Masters and Virginia Johnson, *Human Sexual Inadequacy* (Boston: Little, Brown, 1970), p. 75.
2. Barbara Brown, "Skin Talk: A Strange Mirror of the Mind," *Psychology Today*, August 1974, p. 52.
3. Ashley Montagu, *Touching: The Human Significance of the Skin* (New York: Columbia University Press, 1971), p. 167.
4. Marc H. Hollender, M.D., "The Wish to Be Held," *Archives of General Psychiatry*, vol. 22 (1970), pp. 445–53.
5. Gloria Emerson, "Facing Frank," *The Dial* (Public Broadcasting Corporation publication), November 1981, p. 15.

Chapter 5 (pages 66–80)

1. Desmond Morris, *Manwatching, A Field Guide to Human Behavior* (New York: Harry N. Abrams, 1977), pp. 179–81.
2. Caitlin Thomas, *Not Quite Posthumous Letter to My Daughter* (Boston: Little, Brown, 1963), p. 22.
3. Hendrik Willem Van Loon, *Van Loon's Lives: Napoleon* (New York: Simon and Schuster, 1944), p. 498.
4. Touch for Health Foundation, 1174 N. Lake Avenue, Pasadena, CA 91104.
5. Carlfred Broderick, "Sexual Behavior Among Preadolescents," *Journal of Social Issues*, April 1966, p. 15.
6. David Cole Gordon, *Self-Love* (New York: Verity House, 1968).
7. Karl Menninger, M.D., *Whatever Became of Sin?* (New York: Bantam Books, 1978), pp. 41–42.

Chapter 6 (pages 81–97)

1. Tuan Nguyen, Richard Heslin and Michele Nguyen, "The Meanings of Touch: Sex Differences," *Journal of Communication*, Summer 1975, pp. 92–103; Lawrence Rosenfeld, Sallie Kartus, Ray Chett, "Body Accessibility Revisited," *Journal of Communication*, Summer 1976, p. 27.
2. Shere Hite, *The Hite Report on Male Sexuality* (New York: Knopf, 1981).

3. Suzanne Kurth, "Friendships and Friendly Relations," *Social Relationships* (Chicago: Aldine, 1970), pp. 136–70.
4. Desmond Morris, *Intimate Behaviour* (New York: Bantam Books, 1973), p. 257.
5. Paul Chance, "Facts that Liberated the Gay Community," *Psychology Today*, December 1975, p. 55.

Chapter 7 (pages 98–115)

1. Kenneth J. Gergen, Mary M. Gergen and William H. Barton, "Deviance in the Dark," *Psychology Today*, October 1973, pp. 129–30.
2. Zick Rubin, "Lovers and Other Strangers," *American Scientist*, vol. 62 (March–April 1974), pp. 182–90.
3. Lillian Carter, *Away From Home: Letters to My Family* (New York: Simon and Schuster, 1977), p. 110.
4. Margaret Mary Wood, *The Stranger, A Study in Social Relationships* (New York: Columbia University Press, 1934).
5. Carol Tavris, "The Frozen World of the Familiar Stranger; A Conversation with Stanley Milgram," *Psychology Today*, June 1974, p. 71.

Chapter 8 (pages 116–132)

1. Norman Cousins, *Anatomy of an Illness* (New York: Norton, 1979), p. 154.
2. Elizabeth Hall, "How Cultures Collide: An Interview with Edward T. Hall," *Psychology Today*, June 1976, p. 72.
3. Pamela McCoy, "Further Proof that Touch Speaks Louder than Words," *RN Magazine*, vol, 40, no. 11 (November 1977), pp. 43–46.
4. Lewis R. Wolberg, *The Technique of Psychotherapy* (New York: Grune & Stratton, 1977), p. 514.
5. James R. Petersen, "Eyes Have They But They See Not: An Interview with Rudolf Arnheim," *Psychology Today*, June 1972, p. 72.
6. Dolores Krieger, "Therapeutic Touch: The Imprimatur of Nursing," *American Journal of Nursing*, May 1975, pp. 784–87.

7. Marc H. Hollender and Alexander J. Mercer, "The Wish to be Held and to Hold in Men and Women," *Archives of General Psychiatry*, vol. 33 (January 1976), pp. 49–51.
8. Ashley Montagu, *Touching: The Human Significance of the Skin* (New York: Columbia University Press, 1971), p. 213.
9. Patrick M. O'Neil and Karen S. Calhoun, "Sensory Deficits and Behavioral Deterioration in Senescence," *Journal of Abnormal Psychology*, vol. 84, no. 5 (October 1975), pp. 579–82.
10. Dennis Jaffe, *Healing from Within* (New York: Knopf, 1980).
11. James Roosevelt, *My Parents* (New York: Playboy Press, 1976), p. 113.
12. James Lynch, *The Broken Heart* (New York: Basic Books, 1977), pp. 126–27.

Chapter 9 (pages 133–148)

1. Sonny Kleinfield, *America's Handicapped: The Hidden Minority* (Boston: Little, Brown, 1979), p. 32.
2. Frank Bowe, *Handicapping America: Barriers to Disabled People* (New York: Harper & Row, 1978), pp. 22–23, 126–27.
3. Erving Goffman, *Stigma: Notes on the Management of Spoiled Identity* (Englewood Cliffs, NJ: Prentice-Hall, 1963).
4. Raymond Goldman, *Even the Night* (New York: Macmillan, 1947), pp. 65–66.
5. John Gliedman, "The Wheelchair Rebellion," *Psychology Today*, August 1979, p. 64.
6. Mary Beth Sullivan, Alan H. Brightman and Joseph Blatt, *Feeling Free* (Reading, MA: Addison-Wesley, 1979), p. 156.
7. Irving A. Fein, *Jack Benny: An Intimate Biography* (New York: Pocket Books, 1977), p. 145.

Chapter 10 (pages 149–171)

1. Allan Guttman, *From Ritual to Record: The Nature of Modern Sports* (New York: Columbia University Press, 1978), p. 130.
2. Desmond Morris, *Manwatching: A Field Guide to Human Behavior* (New York: Harry N. Abrams, 1977), p. 305.

3. Andrew Weil, *The Natural Mind* (Boston: Houghton Mifflin, 1972).
4. George Leonard, *The Silent Pulse: A Search for the Perfect Rhythm That Exists in Each of Us* (New York: Dutton, 1978).
5. Richard Grey Sipes, "Sports as a Control for Aggression," *The Humanistic and Mental Health Aspects of Sports, Exercise and Recreation* (Chicago: American Medical Association, 1976), p. 46.
6. "Sex Before Sport?" *Time* magazine, March 15, 1971, p. 54.
7. "Does Sex Sap Strength?" *Family Health* magazine, May 1979, p. 16.
8. David Kopay, with Perry Deane Young, *The David Kopay Story: An Extraordinary Self-Revelation* (New York: Arbor House, 1977), p. 57.
9. William Arens, "The Great American Football Ritual," *The American Dimension: Myths and Social Realities* (Port Washington, NY: Port Washington Press, 1975), pp. 3–14.
10. Paul Koch, *Rip-Off: The Big Game* (New York: Anchor Press/Doubleday, 1972), p. 154.
11. CBS-TV, Los Angeles, Oct. 28, 1978.
12. Guttman, *From Ritual to Record*, p. 53.
13. Harvey Cox, *The Feast of Fools: A Theological Essay on Festival and Fantasy* (Cambridge: Harvard University Press, 1969).
14. Donna Gelfand and Donald P. Hartmann, "Some Detrimental Effects of Competitive Sports on Children's Behavior," *The Humanistic and Mental Health Aspects of Sports, Exercise and Recreation* (Chicago: American Medical Association, 1976), pp. 42–45.
15. Shirley MacLaine, *You Can Get There from Here* (New York: Bantam Books, 1975), p. 121.
16. Andrew Fluegelman, ed., *The New Games Book* (Garden City, NY: Dolphin/Doubleday, 1976).

Chapter 11 (pages 172–187)

1. Sidney B. Simon, "Please Touch: How to Combat Skin Hunger in Our Schools," *Scholastic Teacher Magazine*, Junior/Senior High Teachers' Edition, October 1974, pp. 22–25.
2. Sam Levenson, *You Can Say That Again, Sam!* (New York: Pocket Books, 1975), p. 77.

Chapter 12 (pages 188–209)

1. Lin Farley, *Sexual Shakedown: The Sexual Harassment of Women on the Job* (New York: McGraw-Hill, 1978), p. 39.
2. Robert E. Quinn, "Coping with Cupid," *Administration Science Quarterly*, Cornell University, 1978, pp. 30–45.
3. Roderic Gorney, *The Human Agenda* (West Los Angeles, CA: Guild of Tutors Press, 1979), p. 106.
4. Nancy M. Henley, *Body Politics: Power, Sex and Nonverbal Communication* (Englewood Cliffs, NJ: Prentice-Hall, 1977).
5. Desmond Morris, *Intimate Behaviour* (New York: Bantam Books, 1973), p. 112.

Chapter 13 (pages 210–226)

1. Elizabeth Hall, "How Cultures Collide: An Interview with Edward T. Hall," *Psychology Today*, June 1976, pp. 66.
2. Edward T. Hall, *The Hidden Dimension* (Garden City, NY: Anchor/Doubleday, 1969), p. 120.
3. Edward T. Hall, "The Anthropology of Manners," *Scientific American*, April 1955, p. 85.
4. Hall, *The Hidden Dimension*, p. 150.
5. Elliott Roosevelt, with James Brough, *Mother R.: Eleanor Roosevelt's Untold Story* (New York: G. P. Putnam's Sons, 1977), p. 89.
6. John Money and Herman Musaph, eds., *Handbook of Sexology* (Amsterdam: Elsevier-North Holland Biomedical Press, 1977), p. 1160.
7. Agatha Christie, *Autobiography* (New York: Ballantine, 1977), p. 479.
8. Julie Nixon Eisenhower, *Special People* (New York: Simon and Schuster, 1977), p. 159.

Chapter 14 (pages 227–228)

1. Charles Panati, *Breakthroughs* (Boston: Houghton Mifflin, 1980), p. 6.

BIBLIOGRAPHY

Books

Beisser, Arnold, *The Madness in Sports*. New York: Appleton-Century-Crofts, 1967.

Brown, Barbara, *New Mind, New Body*. New York: Harper & Row, 1974.

Carter, Lillian, *Away from Home: Letters to My Family*. New York: Simon and Schuster, 1977.

Christie, Agatha, *Autobiography*. New York: Ballantine Books, 1977.

Cole, T.M., and S.S. Cole, *The Handicapped and Sexual Health*. New York: Sex Information and Education Council, 1970.

Cousins, Norman, *Anatomy of an Illness*. New York: Norton, 1979.

Cox, Harvey, *The Feast of Fools: A Theological Essay on Festival and Fantasy*. Cambridge: Harvard University Press, 1969.

Critchley, Macdonald, *Language of Gesture*. London: Edward Arnold, 1939.

Eisenhower, Julie Nixon, *Special People*. New York: Simon and Schuster, 1977.

Farley, Lin, *Sexual Shakedown: The Sexual Harassment of Women on the Job*. New York: McGraw-Hill, 1978.

Fleugelman, Andrew, ed., *The New Games Book*. Garden City, NY: Dolphin/Doubleday, 1976.

Fraiberg, Selma, *Every Child's Birthright*. New York: Basic Books, 1977.

Geldard, Frank A., *The Human Senses*. New York: John Wiley, 1972.

Gill, Richard, *White Water and Black Magic*. New York: Holt, 1940.

Goffman, Erving, *Behavior in Public Places*. New York: Free Press, 1963.

———, *Encounters*. Indianapolis: Bobbs-Merrill, 1961.

———, *Stigma: Notes on the Management of Spoiled Identity*. Englewood Cliffs, NJ: Prentice-Hall, 1963.

Goldman, Raymond, *Even the Night*. New York: Macmillan, 1947.

Gordon, Francine E., and Myra H. Strober, eds., *Bringing Women into Management*. New York: McGraw-Hill, 1975.

Gordon, Sol; Peter Scales; and Kathleen Everly, *The Sexual Adolescent*. North Scituate, MA: Duxbury Press, 1979.

Gorney, Roderic, *The Human Agenda*. West Los Angeles, CA: Guild of Tutors Press, 1979.

Grossman, Sebastian P., *A Textbook of Physiological Psychology*. New York: John Wiley, 1967.

Grov, Stanislav, *Realms of the Human Unconscious*. New York: Dutton, 1976.

Guthrie, Russell Dale, *Body Hot Spots*. New York: Van Nostrand, Reinhold, 1976.

Hall, Edward T., *Beyond Culture*. Garden City, NY: Anchor/Doubleday, 1977.

———, *The Hidden Dimension*. Garden City, NY: Anchor/Doubleday, 1969.

———, *The Silent Language*. Garden City, NY: Anchor/Doubleday, 1973.

Hammond, Sally, *We Are All Healers*. New York: Harper & Row, 1973.

Hartman, William E., and Marilyn A. Fithian, *Treatment of Sexual Dysfunction, A Bio-Psycho-Social Approach*. Long Beach, CA: Center for Marital and Sexual Studies, 1972.

Hendin, Herbert, *The Age of Sensation*. New York: Norton, 1975.

Henley, Nancy M., *Body Politics: Power, Sex and Nonverbal Communication*. Englewood Cliffs, NJ: Prentice-Hall, 1977.

Howells, William, *Mankind in the Making*. Garden City, NY: Doubleday, 1967.

Jonas, Gerald, *Visceral Learning*. New York: Viking Press, 1973.

Kleinfield, Sonny, *America's Handicapped, The Hidden Minority*. Boston: Little, Brown, 1979.

Kopay, David, and Perry Deane Young, *The David Kopay Story*. New York: Arbor House, 1977.

Krieger, Dolores, *Therapeutic Touch*. Englewood Cliffs, NJ: Prentice-Hall, 1979.

Leboyer, Frederic, *Birth Without Violence*. New York: Knopf, 1976.

Leonard, George, *The Ultimate Athlete*, New York: Viking Press, 1975.

Lifchez, Raymond, and Barbara Winslow, *Design for Independent Living*. New York: Watson-Guptill, 1979.

Lynch, James, *The Broken Heart*. New York: Basic Books, 1977.

MacLaine, Shirley, *You Can Get There from Here*. New York: Bantam, 1975.

Mark, Vernon H., and Frank R. Ervin, *Violence and the Brain*. New York: Harper & Row, 1970.

McCall, George J., *Social Relationships*. Chicago: Aldine, 1970.

McKuen, Rod, *Alone*. New York: Pocket Books, 1975.

Menninger, Karl, *Whatever Became of Sin?* New York: Bantam, 1978.

Michener, James A., *Sports in America*. New York: Random House, 1976.

Milgram, Stanley, *Obedience to Authority*. New York: Harper & Row, 1974.

Montagu, Ashley, *Touching: The Human Significance of the Skin*. New York: Columbia University Press, 1971.

Morris, Desmond, *Intimate Behavior*. New York: Bantam, 1973.

———, *Manwatching: A Field Guide to Human Behavior*. New York: Harry N. Abrams, 1977.

Murray, Henry, *Human Nature in Politics*. Oxford, England: Oxford University Press, 1938.

Orlick, Terry, *Winning Through Cooperation*. Washington, DC: Acropolis Books, 1978.

Packard, Vance, *A Nation of Strangers.* New York: Pocket Books, 1974.

Panati, Charles, *Breakthroughs.* Boston: Houghton Mifflin, 1980.

Penfield, Wilder, *The Mystery of the Mind.* Princeton, NJ: Princeton University Press, 1975.

Rabin, Barry J., *The Sensuous Wheeler.* San Francisco: Multi Media Resource Center, 1980.

Rights of Physically Handicapped People, The, ACLU Handbook. New York: Discus/Avon, 1979.

Roosevelt, Elliott, with James Brough, *Mother R.: Eleanor Roosevelt's Untold Story.* New York: G. P. Putnam's Sons, 1977.

Sagn, Carl, *Dragons of Eden.* New York: Ballantine Books, 1977.

Simon, Sidney B., *Caring, Feeling, Touching.* Niles, IL: Argus Communications, 1976.

Snyder, Eldon E., and Elmer Spreitzer, *Social Aspects of Sport.* Englewood Cliffs, NJ: Prentice-Hall, 1978.

Sorell, W., *Story of the Human Hand.* London: Weidenfeld & Nicolson, 1968.

Spitz, René, *The First Year of Life.* New York: International Universities Press, 1965.

Sullivan, Mary Beth; Alan J. Brightman; and Blatt, Joseph, *Feeling Free.* Reading, MA: Addison-Wesley, 1979.

Thie, John F., with Mary Marks, *Touch for Health.* Pasadena, CA: Touch for Health Foundation, 1973.

Tillich, Hannah, *From Time to Time.* New York: Stein & Day, 1973.

Ullmann, Liv, *Changing.* New York: Knopf, 1977.

Van Lawick-Goodall, Jane, *In The Shadow of Man.* Boston: Houghton Mifflin, 1971.

Vernon, Jack, *Inside the Black Room: Studies of Sensory Deprivation.* New York: Clarkson N. Potter, 1963.

Weatherhead, Leslie D., *Psychology, Religion and Healing.* Nashville, TN: Abingdon-Cokesbury Press, 1951.

Weil, Andrew, *The Natural Mind.* Boston: Houghton Mifflin, 1972.

Wilde, Meta Carpenter, *A Loving Gentleman.* New York: Simon and Schuster, 1976.

Wolberg, Lewis R., *The Technique of Psychotherapy.* New York: Grune & Stratton, 1977.

Wolfe, Charlotte, *The Hand in Psychological Diagnosis*. New York: Philosophical Library, 1952.

Wood, Margaret Mary, *The Stranger, A Study in Social Relationships*. New York: Columbia University Press, 1934.

Wright, Beatrice A., *Physical Disability, A Psychological Approach*. New York: Harper & Row, 1960.

Zubek, John Peter, *Sensory Deprivation*. New York: Appleton-Century-Crofts, 1969.

Articles, Monographs, Unsigned Articles

Barber, Phyllis, "Touching," *Utah Holiday Magazine*, June 1980, p. 26.

Bayley, Nancy, "Sensory Development," *International Encyclopedia of the Social Sciences*, vol. 14. New York: Macmillan, 1968, p. 183.

Blackman, Nancy B., "Pleasure and Touching: Their Significance in the Development of the Preschool Child," Paper presented at the International Symposium on Childhood and Sexuality, Montreal, September 1979.

Boll, Thomas J.; Berent, Stanley; and Richards, Herbert, "Tactile-Perceptual Functioning as a Factor in General Psychological Abilities," *Perceptual and Motor Skills*, April 1977, pp. 535–39.

Breed, George, and Ricci, Joseph S., "Touch Me, Like Me," In *Annual Proceedings, American Psychological Association*, vol. 8 (1973), p. 153.

Burton, Arthur, and Heller, Louis G., "The Touching of the Body," *Psychoanalytic Review*, vol. 51, no. 1 (Spring 1964), pp. 122–34.

Cannon, Rose Broeckel, "The Development of Maternal Touch During Early Mother-Infant Interaction," *JOGN Nursing*, vol. 6, no. 2 (March/April 1977), pp. 28–33.

Chodoff, Paul, "The Seductive Patient," *Medical Aspects of Human Sexuality*, vol. 2, no. 2 (February 1968), p. 52.

Clay, Vidal S., "The Effect of Culture on Mother-Child Tactile Communication," *Family Coordinator*, vol. 17 (1968), pp. 204–10.

Coppola, August, "Reality and the Haptic World," *Phi Kappa Phi Journal*, Winter 1970, p. 14.

Cowan, Ed, "Why Sport?" *The Humanist Magazine*, (November/December 1979), p. 25.

DeRosis, Helen A., "Violence, Where Does It Begin?" *The Family Coordinator*, National Council on Family Relations, October 1971, p. 361.

Dies, Robert R., and Greenberg, Barbara, "Effect of Physical Contact in an Encounter Group Context," *Journal of Consulting and Clinical Psychology*, vol. 44, no. 3 (June 1976), pp. 400–05.

Duke, Marshall P., et al. "Comfortable Interpersonal Distance," *Journal of Experimental Research in Personality*, vol. 6 (1972), pp. 119–32.

Dundes, Alan, "Into the Endzone for a Touchdown; A Psychoanalytic Consideration of American Football," *Western Folklore*, vol. 37 (1978), pp. 75–88.

Endsley, Richard C., "Effects of Maternal Model on Child Tactual Curiosity," *Journal of Genetic Psychology*, vol. 131 (September 1977), pp. 21–28.

Fisher, Jeffrey D.; Rytting, Marvin; and Heslin, Richard, "Hands Touching Hands," *Sociometry* (December 1976), p. 416.

Frank, Lawrence, K., "Tactile Communication," *Genetic Psychology Monographs*, vol. 56 (1957), pp. 209–55.

"Friendly Touch," unsigned article, *Human Behavior Magazine*, January 1977, p. 47.

Gelfand, Donna M., and Hartmann, Donald P., "Some Detrimental Effects of Competitive Sports on Children's Behavior," in *The Humanistic and Mental Health Aspects of Sports, Exercise, and Recreation*, Chicago: American Medical Association, 1976.

Gibson, James J., "Observations on Active Touch," *Psychological Review*, vol. 69, no. 6 (November 1962), pp. 477–91.

Goffman, Erving, "The Nature of Deference and Demeanor," *American Anthropologist*, vol. 58 (1956), pp. 473–502.

Hall, Edward T., "Psychological Aspects of Foreign Policy," *Proceedings before Committee on Foreign Relations, U.S. Senate*, June 5, 19, 20, 1969.

Hall, Elizabeth, "How Cultures Collide: An Interview with Edward T. Hall," *Psychology Today*, June 1976, p. 72.

Heslin, Richard, "Steps Toward a Taxonomy of Touching," paper presented at Midwestern Psychological Association meeting, Chicago, May 1974.

———, and Boss, Diane, "Nonverbal Intimacy in Arrival and Departure at an Airport," paper presented at Midwestern Psychological Association meeting, Chicago, 1977.

Hollender, Marc. H., "The Wish To Be Held," *Archives of General Psychiatry*, vol. 22 (1970), pp. 445–53.

Hollender, Marc. H. and Alexander J. Mercer, "The Wish to be Held and to Hold in Men and Women," *Archives of General Psychiatry*, vol. 33 (January 1976), pp. 49–51.

"How People React to Your Touch," *Science Digest*, March 1976, p. 46.

Jennings, Lane, "Future Fun; Tomorrow's Sports and Games," *The Futurist Magazine*, World Future Society, vol. XIII, no. 6 (December 1979), p. 417.

Kleck, Robert, "Physical Appearance Clues and Interpersonal Attraction in Children," *Child Development*, vol. 45 (1974), pp. 305–10.

———, "Stigma as a Factor in Social Interaction," in *Perspectives on Human Deprivation: Biological, Psychological, and Sociological*, NICHD Task Force Report, Washington, DC: U.S. Department of Health, Education, and Welfare, 1968.

———, et al, "Effect of Stigmatizing Conditions on the Use of Personal Space," *Psychological Reports*, vol. 23 (1968), pp. 111–18.

Ladieu, G., "Social Acceptance of the Injured," *Journal of Social Issues*, vol. 4, no. 4 (1948), pp. 55–61.

Larue, Gerald A., "Grief," *The Humanist Magazine*, July/August 1978, p. 4.

Lown, Bernard and Segal, Jack, "Post Myocardial Infarction Care: How to Manage Your Patient's Arrhythmias," *Modern Medicine Magazine*, Sept. 30–Oct. 15, 1978, pp. 60–77.

Major, Brenda, and Heslin, Richard, "Perceptions of Same-Sex and Cross-Sex Touching," paper presented at Midwestern Psychological Association Conference, Chicago, 1978.

McCoy, Pamela, "Further Proof that Touch Speaks Louder Than Words," *RN Magazine*, vol. 40, no. 11 (November 1977), pp. 43–46.

Musaph, Herman, "Skin, touch and sex," *Handbook of Sexology*, John Money and Herman Musaph, eds. Amsterdam: Elsevier/North-Holland Biomedical Press, 1977, p. 1160.

Ohno, Mary I., "The Eye-Patched Patient," *American Journal of Nursing*, vol. 71, no. 2 (February 1971), p. 271.

Olds, James, "Pleasure Centers in the Brain," *Scientific American*, vol. 195 (October 1956), pp. 105–17.

O'Neil, Patrick M. and Karen S. Calhoun, "Sensory Deficits and Behavioral Deterioration in Senescence," *Journal of Abnormal Psychology*, vol. 84, no. 5 (October 1975), pp. 579–82.

Pattison, Joyce E., "Effects of Touch on Self-Exploration and the Therapeutic Relationship," *Journal of Consulting and Clinical Psychology*, vol. 40, no. 2 (1973), pp. 170–75.

Petersen, James R., "Eyes Have They, But They See Not: An Interview with Rudolf Arnheim," *Psychology Today*, June 1972, p. 72.

Prescott, James W., "Body Pleasure and the Origins of Violence," *The Futurist Magazine*, April 1975, pp. 64–73.

Rohrbaugh, Joanna Bunker, "Femininity on the Line," *Psychology Today*, August 1979, p. 30.

Roiphe, Anne, "The Lost Art of Touching," *Ladies' Home Journal*, May 1977, p. 105.

Rose, Susan A.; Schmidt, Katalin; and Bridger, Wagner H., "Cardiac and Behavioral Responsivity to Tactile Stimulation in Premature and Full-Term Infants," *Developmental Psychology*, vol. 12, no. 4 (1976), pp. 311–20.

Rubin, Zick, "Lovers and Other Strangers," *American Scientist*, vol. 62 (March/April 1974), pp. 182–90.

Rynearson, Robert R., "Touching People," *Journal of Clinical Psychiatry*, vol. 39, no. 6 (June 1978), p. 492.

Sandroff, Ronni, "A Skeptic's Guide to Therapeutic Touch," *RN Magazine*, vol. 43, no. 1 (January 1980), p. 25.

Simon, Sidney B., "Please Touch: How to Combat Skin Hunger in Our Schools," *Scholastic Teacher Magazine*, Junior/Senior High Teachers' Edition, October 1974, pp. 22–25.

Sipes, Richard Grey, "Sports as a Control for Aggression," in *The*

Humanistic and Mental Health Aspects of Sports, Exercise, and Recreation, Chicago: American Medical Association, 1976.

"Size Discrimination on Skin," unsigned article in *Science Magazine*, Jan. 31, 1969, p. 488.

Snyder, Melvin L., et al., "Avoidance of the Handicapped," *Journal of Personality and Social Psychology*, vol. 37, no. 12 (1979), pp. 2297–2306.

"Social Environment and Brain Chemistry," unsigned article in *American Association for the Advancement of Science Bulletin*, November 1966.

Sommer, R., "Studies in Personal Space," *Sociometry*, vol. 22 (1959), pp. 247–60.

Tolor, A., "Psychological Distance in Disturbed and Normal Adults," *Journal of Clinical Psychology*, vol. 26 (1970), pp. 160–62.

Turner, Edward T., "The Effects of Viewing College Football, Basketball and Wrestling on the Elicited Aggressive Responses of Male Spectators," in *Contemporary Psychology of Sports; Proceedings of the Second International Congress of Sport Psychology, Washington, DC, 1968* (Madison, WI: University of Wisconsin, 1968).

Wolfgang, Marvin, E., "Violence and Human Behavior," in *Proceedings of the Annual Conference of the American Psychological Association*, Washington, DC, Aug. 30, 1969.

INDEX